100

LIVESTREAMING & DIGITAL MEDIA PREDICTIONS

VOLUME 4

*Top Content Creators Help You Succeed Online
in an Era of Rapid Change*

ROSS BRAND

livestreamuniverse.com

THE 100 PREDICTIONS SERIES

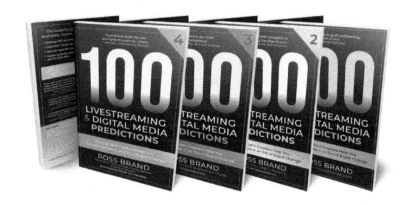

#1 Amazon Best Seller in Six Countries

Winner of 25 Book Awards

JOIN OUR UNIVERSE

As a valued *100 Predictions* reader, you are invited — and encouraged — to become a part of our Livestream Universe community. Look for new content to keep you informed and entertained and new initiatives to help grow your influence and monetize your content.

The best way to stay in the loop is to join our free email list: **livestreamuniverse.com/streamleader**

The best way to get our latest content is to subscribe to our YouTube channel:

youtube.com/@livestreamuniverse

The best way to connect with the book's author and contributors is to join our free Facebook Group:

facebook.com/groups/livestreamuniversegroup

For my parents, Marion & Irv Brand

REACTIONS
TO VOLUME 3

"I know I say this every year, but this resource delivers on this promise every time: if you want to learn how to get far more visible and make a much better impact through digital media and livestreaming, this book series is the only resource you need. Yes, my opinion is a wee-bit biased given that I'm featured inside - but I have to say that your Predictions books are AMAZING, Ross! Volume 3 packs an even better punch given the volatile nature of the digital landscape these days. Like the annual blog series that preceded the books (which I was also proud to be featured in) I love how Ross's series provides a fun, super quick, easy to read, and most importantly, ACTIONABLE collection of predictions from a wide variety of livestreamers and digital media geniuses. Think of it as part cheat sheet to this online world and part rolodex of digital badassery. Each contributor offers a unique perspective for you to consider as you press on with your own digital media endeavors. It's the Who's Who of our space and time, and I am truly honored that Ross keeps inviting my perspective back to be included among such superstars - with Ross himself chief among us!"

-Jennie Mustafa-Julock aka Coach Jennie

"In Volume 3 of 100 Livestreaming & Digital Media Predictions, Ross Brand (and friends) have provided an easy-to-digest collection of insightful tips sure to help anyone navigating not just the livestream world, but digital marketing and tools (including ChatGPT) as a whole. Even more important, this book—like its predecessors—is full of quick wins sure to shorten the learning curves of everyone from novice to expert. Well done, Ross!"

-Mark Babbitt, co-author of Good Comes First

"Ross Brand has done it again, gathering the latest thoughts on digital media and live-streaming from many of the leaders in the space. I learn a lot not just because there are people here smarter than me (there are) but because they all bring their A-game for Ross's book."

-Doug Cohen, M10Social

"This book is a collection of predictions from top content creators in the field of livestreaming and digital media. It provides valuable insights into the future of these industries and offers strategies for success in an era of rapid change. Ross Brand has compiled a diverse group of experts who provide their perspectives on a wide range of topics, including the use of new technologies, the impact of social media on livestreaming, and the future of video production. This book is a great resource for anyone looking to stay ahead of the curve in the ever-evolving world of digital media and livestreaming.."

-Brad Friedman, author of The Small Business Owner's Guide to Inbound Marketing

"Once again Ross and friends knocked it out of the park by providing quicks and easy wisdom and strategy for navigating the live video and broader digital marketing arena. From timely information on ChatGPT to trends in live streaming and live selling, this is very valuable to hear from so many leaders in the industry. It's like attending a conference and hearing from 100+ different speakers!"

-Mike Gingerich, Content Marketer

"Let it guide you through the livestreaming world! LOTS and lots of great insight is offered in this book! There's always room for growth in the world of Livestreaming and this book will guide you. Ross Brand and all the contributors deliver clean and concise predictions that can't be found elsewhere!"

-Marisa Cali, Virtual Event Producer

"100 Livestreaming & Digital Media Predictions, Volume 3: Top Content Creators Help You Succeed in an Era of Rapid Change is a book that spotlights livestreaming and digital media predictions form some of the top content creators and digital entrepreneurs globally. Each chapter has a stand-alone prediction that authentically shares their insights on what to expect in the arena as we head into 2023. With each prediction I have gained more clarity on what to expect next, how to maneuver the digital world both personally and professionally as I head into 2023. I highly recommend this book."

-Joie Gharrity, Superstar Women Entrepreneurs Network

"Livestreaming will impact your business, your buying, your learning, and your recruiting! Volume 3 of Ross Brand's Livestreaming predictions series is a fast read full of recommendations for livestreaming & digital media apps, tools, equipment, and - most importantly - likely trends that will impact you this year. Whether you're an established streamer or just considering trying livestreaming, you'll find reliable tools and practices that are immediately of value."

-Chris Edmonds, co-author of *Good Comes First*

"Clear, concise advice about the digital future to come. Ross has done it again! Reading through the many predictions, I was struck by how much things had changed in only a year. Really interesting sections in this book about AI, Web3, vertical video, social media, building communities, and a lot more. Reading through it, I learned a great deal of actionable tidbits and I look forward to going through it again for more. If you want to know the future of livestreaming and digital media productions, you can't go wrong by picking up this book. It'll definitely give you a head start on the year to come!"

-Jodi Krangle, voice actor and audio branding expert

"This is a book that's not just predictions, but a guide on how to create content that matters in the digital world we live in. It's a must-read for anyone looking to stand out in the crowded digital space. Here's the deal- In "100 Livestreaming & Digital Media Predictions," the authors expertly address the challenge of creating effective and engaging content in today's fast-paced digital world. They highlight the issue of the overwhelming amount of content that is produced and consumed daily, making it increasingly difficult for creators to cut through the noise and connect with their audience. The good news is that throughout each chapter, the authors offer unique solutions to the above by providing actionable tips and predictions for how to stay ahead of the curve and create content that truly resonates with viewers. From leveraging audio and video livestreaming and social media platforms to understanding the latest trends in digital media including web3, the metaverse and AI, this book is a must-read for anyone looking to up their content creation game.

-Mitch Jackson, 2013 California Litigation Lawyer of the Year & Co-Founder ManeuVR

"Must Read to Stay Ahead of the Online Trends! This book features so many talented live streamers & digital marketing experts sharing their predictions for 2023. You will want to check out every chapter for all the insights & what to expect in 2023 in a rapidly changing industry."

-Erin Cell, Socially Powered & The NiFTy Chicks Co-Host

"Wow! What a great collection of expert insights! Ross Brand brings together such a great host of experts from so many different angles of digital media and marketing. I learned a ton from the different predictions and tips. I also enjoyed the short sections for each, making it easy to hop around the book as I wanted."

-Jesse Randal

"Good resource to help you start livestreaming. What I love about the book are the great resources at the end. I've started livestreaming daily because of the encouragement from Ross. I encourage you to pick up the book today."

-Meiko S. Patton, AI Newsletter Strategist

"Can't get this info anywhere else! This book contains more than just helpful predictions from world-class professionals; it contains top level tips, hard-earned wisdom, expert visions of the future, deep insights and more! Staying "current" in the rapidly changing world of digital media is not easy but this book HELPS. SO. MUCH."

-Amazon Customer

Predictions that will make you think! Volume 3 of 100 Live-streaming & Digital Media Predictions definitely gives you some great insights as to the changes coming in the content creator and digital marketing space. A great read for anyone looking to figure out where to focus their energy on learning and how to connect with leaders in the field. Ross Brand has done a great job of putting together a great resource for all.

-Jim Fuhs, Dealcasters Live Co-Host

"What can you get when you have so many great professionals featured in the book? Great insights, diversity, and a great shortcut to start or improve your livestream, content creation, digital marketing, or entrepreneur journey. You can follow the order or randomly navigate through the pages, ending with lots of powerful tools."

-Marco Novo, Amazon Live Influencer

Forge your unique voice, build a die-hard community, produce killer content, align with the right tribes and never compromise on quality. Do this, and not only will you survive in this hyper-competitive arena — you'll thrive!

-Phil Gerbyshak

The future of digital media will be shaped by those who can seamlessly integrate AI tools into their strategy without losing the human touch.

-Louise McDonnell

Any opportunity to show that your brand is genuine, ethical and customer/client-centric should be built into your marketing strategies.

-Karen Graves

If we embrace emerging technologies with wisdom, foresight and care, AI can take content creation to thrilling new heights.

-Christoph Trappe

TABLE OF CONTENTS

FOREWORD BY *Rob Greenlee, Claudia Santiago & Rebecca Gunter*.................. 1

PART I 6

THE BIG PICTURE 6

INTRODUCTION 7

PART II 19

PREDICTIONS 19

1. Chetachi A. Egwu, Ph.D. 20

2. Rob Greenlee 23

3. Clarice Lin 25

4. Mitch Jackson 27

5. Nicole Sanchez 29

6. Larry Roberts 31

7. Jodi Krangle 33

8. Mike Allton 35

9. Judi Fox 38

10. Dale L. Roberts 40

11. Valerie Morris 43

12. Roger Wakefield 45

13. Claudine Francois 47

14. Nick Nimmin 50

15. Junaid Ahmed 52

16. JennyQ 55

17. Jon Burk 57

18. Marisa Cali 59

19. Jeff Sieh 61

20. Brian Schulman 62

21. Claudia Santiago 65

22. Rebecca Gunter 67

23. Dan Currier 68

24. Jeffrey Powers 70

25. Lee Uehara 72

26. Liron Segev 74

27. Dave Jackson 76

28. Caren Glasser 78

29. Brenden Mulligan 80

30. Toni Henderson-Mayers 82

31. Brad Friedman 83

32. S. Chris Edmonds 86

33. Amir Zonozi 87

34. Kerstin Oleta 89

35. Ross Quintana 91

36. Bryan Kramer 93

37. Renee Hastings 94

38. Andrew Kavanagh 97

39. Professor Nez 99

40. Nancy Debra Barrows 100

41. Dr. Aikyna Finch 104

42. Chad Illa-Petersen 106

43. Kevin Kolbe 108

44. Jennie Mustafa-Julock 109

45. Christoph Trappe 111

46. Alessandra Colaci 114

47. Phil Gerbyshak 116

48. Niel Guilarte 119

49. Joie Gharrity 120

50. John Giovanni Pretto 121

51. Lottie Hearn 124

52. Desiree Duffy 126

53. Meiko S. Patton 128

54. Chris Stone 129

55. Bridgetti Lim Banda 133

56. Chris Krimitsos 136

57. Jim Fuhs 137

58. Elikqitie 139

59. Marc Gawith 140

60. John Largent 142

61. Brian Wallace 144

62. Bee Smith 146

63. Eric Hunley 147

64. Karen Yankovich 149

65. Paul Richards 151

66. Katarina Andersson 152

67. Laura Clapp Davidson 154

68. Win Charles 155

69. Dr. Stuart Buchan 156

70. Katie Hornor 158

71. Terry Brock 159

72. Karen Graves 160

73. Jeffrey Fitzgerald 162

74. Beth Granger 166

75. Wágner dos Santos 167

76. Jessica Kupferman 170

77. Chris Curran 171

78. Louise McDonnell 172

79. Jonathan Tripp 174

80. John Kapos 175

81. Terry Johnson 177

82. Melanie Falvey 179

83. Lona DeRieux 182

84. JS Gilbert 184

85. Nancy Myrland 186

86. Mike Gingerich 188

87. Sue-Ann Bubacz 190

88. Marco Novo 194

89. Jaime Legagneur 197

90. Mary Barnett 198

91. Julia Jornsay-Silverberg 202

92. Jeff Howell 203

93. Doug Cohen 205

94. Jaime Cohen 207

95. Mario Fachini 208

96. Phil Kluba..211

97. Kyle M. Bondo 212

98. Tamara Thompson 216

99. Vicki Fitch...217

100. Doyle Buehler..219

101. Janine Nicole Dennis.................................221

102. Timothy Kimo Brien 222

103. Rob Cairns 223

104. Nicole Christina 224

105. Zack Bordeaux...225

106. Christine Gritmon 226

107. Tim Gillette...228

108. Anita Sonya 229

109. Carlos Phoenix Jimenez 231

110. Steven Healey 233

111. Gord Isman 235

112. Ian Anderson Gray.....................................237

113. Stephanie Garcia 239

PART III 243

RESOURCES 243

NEXT STEPS 244

WHAT'S IN ROSS' CREATOR TOOLKIT?................245

NEW ADDITIONS FOR VOLUME 4 246

LIVESTREAMING & VIDEO PRODUCTION SOFTWARE....................249

PODCAST SOFTWARE.....................................252

AI TOOLS 253

AUDIO & VIDEO GEAR....................................254

GRAPHICS & SHORT VIDEOS 258

WEBSITE, EMAIL & MARKETING 259

GLOSSARY 261

ACKNOWLEDGEMENT 266

ABOUT THE AUTHOR 268

FOREWORD BY
ROB GREENLEE, CLAUDIA SANTIAGO & REBECCA GUNTER

...with love from the StreamLeader Report LIVE Panel people

I t isn't easy to shine online, but you don't have to do it alone. That's what we've learned from watching our man Ross Brand deftly engage an audience, the comment section and us, his quirky group of weekly co-hosts on the **StreamLeader Report** YouTube LIVE panel show.

For you see, despite the longevity, reputation and *Podcaster Hall of Fame*-status of **Rob Greenlee**, the energized audiences that enthusiastically support world-renowned entertainer **Claudia Santiago**, and the unique POV of earnest newcomer-turned-avid creator and contributor **Rebecca Gunter**, when it comes to bringing distinct voices together seamlessly, week after week, in a quality show with something to say, Ross is the G.O.A.T. — his unfettered enthusiasm for broadcasting, livestreaming and digital media is so genuine you can almost taste it. The man has earned his reputation for being an industry innovator and big picture thinker.

As such, when Ross asked us to come together and co-author the foreword for *100 Livestreaming & Digital Media Predictions, Vol 4,* we couldn't say "yes" fast enough. It's one thing we all agree on: In Ross' world, **the future is 100.**

Don't just take our word for it, friends, because for four years in a row, one hundred industry pros, content creators and small business owners have heeded Ross's call to action and offered their opinion: *What lies ahead for livestreaming and digital media?*

By collecting and curating prognostications of a well-nurtured and many-splendored community of creators, Ross Brand's '100 Predictions' has been serving up annual industry hot takes since 2016.

These are the voices, perspectives and predictions — straight outta the zeitgeist — that futurecasts a world where business, brands and creative thought-leaders use their voice to deepen their unique value(s) and cultivate an audience by embracing the tech, tools and trajectory of platform-building and digital media.

If you're new here, been a fan of this series, or are even one of the three contributors who have contributed every single year, <tips hat>, welcome. (We see you Mitch Jackson, Jennie Mustafa-Julock, and JennyQ!)

We're so excited to introduce you to *100 Livestreaming & Digital Media Predictions, Vol 4* **because this edition is extraordinary.**

In this installment, creators of every stripe encourage us all to deepen our relationships and experiences with livestreaming and each other. Want a sneak peek into the hive mind of this year's most compelling ideas? Small and regional events take center stage in the world of digital media, exploring the value of smaller, more intimate audiences. Speaking of small, micro-content and micro-learning are very much in fashion for 2024. And it should come to no one's surprise that public opinion of AI is the hot topic du jour — is this irrefutable technology friend or faux?

"Each prediction feels like a full chapter," Ross shares with us excitedly, "the quality of this year's submissions are <chef's kiss>." The vibe is collaborative and considered. Folks from all walks of life have shown up to offer their perspectives and predictions. This is a project contributors are proud to be a part of.

Ross Brand Encourages Us All to "Stream Bigger"

If you know Ross at all, it's easy to see why 100 people put so much thoughtful consideration into their predictions to bring this book together. Since the very first day he clicked "go live" in 2015, Ross has made it his mission to show up for people. His reputation precedes him in profound ways — this is a man who wants the best for people who are willing to do what it takes to be a creator, a true cultivator of win-win scenarios and champion for rising tides that lift all boats.

In many ways, the annual co-creation of *100 Livestreaming & Digital Media Predictions* is our way of showing up for him, the person who has always shown up for us. If you've evolved as a creator, grown through mentorship, felt grounded and supported, learned how to grow an audience, well, it's more than likely that Ross Brand had something to do with it, even tangentially.

100 Experts Crystal Ballin'

In the true spirit of his entrepreneurial energy and community building, as a result you are holding in your hands not just another crowd-sourced echo chamber, this, dear reader, is the convergence of platforms, people and POV that you didn't even know you needed.

Browse a curated catalog of ideas, inspiration and unignorable advice from the who's who of livestreaming and digital media. Explore trends, topics and techniques from the people who embrace them. Join the conversation here at 'Co-Creation Nation,' a community of creators united in a common goal: Be. More. Visible.

Collectively, *100 Livestreaming & Digital Media Predictions, Vol 4* is your one-stop shop for peer-to-peer support for content creators and media makers. Expert-led and community-driven, these predictions offer cautionary tales and life lessons from livestreamers, podcasters and YouTubers that will help content creators at any level navigate rapid change... with a little help from your 100 new best friends.

Crystal Ball Meets Time Capsule, Voices of the Digital Age

What is timeless is in equal measure with what is changing. And thus, we contend that this series not only serves as a valuable resource right now, it is also tomorrow's history. If "Broadcasting and Content Creation in the Digital Age" becomes a class in the future, this is the book.

What did the action takers, quality conversation starters, and leaders with integrity do in 2024 to make media matter more than just collecting eyeballs and pleasing advertising/algorithms? What can we learn about the moment in time when many mediums — podcasting, video podcasting, livestreaming, broadcasting — converged in a cacophony of emerging technologies?

Understanding where we've been to understand where we're going, because context is king.

Evolutionary > Revolutionary

100 Livestreaming & Digital Media Predictions, Vol 4 is a must-read book for anyone serious about impacting the digital media content creation world. This new edition is full of real-world tips and insights from the content creator economy leaders — a treasure trove of innovative strategies, trends and monetization tips. It's not just a guide, it's the key to unlocking your potential in the rapidly evolving landscape of online content creation. Tech, tools, and teachable moments, this series is a true game-changer for any aspiring digital media — enthusiasts and professionals alike — who could benefit from 'SME (subject-matter-expert) Cheat Codes,' year after year.

100 POV | 1 Place

And so, without further adieu, turn the page to deepen your relationships and experiences with other industry people, content creators and platform builders who predict a future where business, brands and creative thought-leaders use their voice to deepen their unique value(s) and cultivate an audience — because it isn't easy to shine online, but you don't have to do it alone. #StreamBigger

With love from the StreamLeader Report LIVE Panel people and forever friends to the community of creators we are happy and honored to be a part of,

Rob Greenlee, Claudia Santiago & Rebecca Gunter
Co-hosts, StreamLeader Report LIVE Panel

PART I
THE BIG PICTURE

INTRODUCTION

In the 2024 digital media space, the pace of change can seem dizzying. Even if you're an early adopter, you might feel overwhelmed by all the shiny new AI (artificial intelligence) objects seemingly introduced daily. You want to tune out some of the noise and focus on what you do best, yet you fear falling behind. You don't want to lose out to your colleagues and competitors who are raving about the newest AI integration that saves them hours of work while enabling them to massively increase their productivity.

If you are new to digital media creation, then a roadmap to this complex and ever-changing online universe would sure come in handy.

Fortunately, our 113 expert contributors share their insights on not only what will happen, but how best to deal with the challenges we all face in this new era of AI and digital media. Those challenges aren't small, and the way you adapt today can impact how your business performs — and if it survives — in 2024 and beyond.

If this is the first volume of this series instead of the fourth, we might call it "100 AI & Digital Media Predictions," given the prominence of AI throughout this year's entries.

Well, probably not.

While many predictions include AI, they aren't only about AI. The book's contributors recognize AI as a tool that can help readers create faster and with higher quality. It's about AI serving as one solution

— not the only solution — to the challenges faced by livestreamers, podcasters, YouTubers, email marketers and community builders.

Although opinions differ on whether AI is a net positive or negative for humanity, the consensus view is that for creators, AI is your helper, not a ticket to irrelevance.

Kevin Kolbe (Chapter 43), a YouTuber and best-selling author, astutely notes, "AI can be a great tool for ideas and curation, but the irony is AI still needs a human to tell it what to do." This sentiment underscores a crucial theme of this book: AI as an extension of human creativity.

AI is coming at us from all directions, resulting in a whole slew of new products and added features to the software staples in our established workflows. We need to decide how much time to invest in learning about, testing out and integrating AI into our business and media production processes.

Will it hurt your content and competitiveness if you offload tasks to AI? Or will it open opportunities to create a volume of media previously unthinkable without a big team or organization behind you?

One concept the thought-leaders featured in the upcoming chapters largely agree on is that the more AI pushes its way into our lives, the more important it is to bring YOU to the content you create and the people you interact with. "We're going to continue to see a huge shift to AI streamlining and infiltrating almost anything we do," says Valerie Morris (Chapter 11), a social media and digital marketing strategist. "That's only going to make community and authentic humanity more valuable."

She says to put AI to work on the productivity side but stay true to your authentic self when it comes to human interaction. "Yes, AI has a huge opportunity in marketing and communications, but there's something special about real-time, authentic interaction and insight," Morris adds.

While acknowledging the pros and cons of AI, YouTuber and Amazon Influencer Dan Currier (Chapter 23) agrees with Morris that AI helps work get done faster and more efficiently. "The creator community cannot overlook the raw potential AI represents in providing new levels of efficiency to our workflows, that would otherwise be impossible," says Currier.

He advises creators not to sit on the sidelines in 2024. "You don't have to learn EVERYTHING about AI all at once," Currier says. "Try something simple like having ChatGPT write your social media profile or video descriptions. Whether you choose to dive headfirst or just dip your toe, this is the year to get started."

The book's contributors help us navigate a landscape full of opportunities, even as they are point out the pitfalls to watch out for. Talk-show host Caren Glasser's (Chapter 28) perspective brings to light another pain point: The struggle for smaller creators and businesses to find their footing in a landscape often dominated by larger players. She sees AI's leveling the playing field and enabling independent voices to be heard and valued.

Meanwhile, tech YouTuber Liron Segev (Chapter 26) warns of its darker applications by malevolent actors, including YouTube channel hackers and email scam artists. Ethical considerations rise to the fore as Hall of Fame speaker Terry Brock (Chapter 71), content strategist

Christoph Trappe (Chapter 45) and livestreaming OG John Giovanni Pretto (Chapter 50) highlight the need for a vigilant and principled approach to AI's advancement, ensuring its alignment with humanity's best interests.

The final word on AI for this section of the introduction goes to Claudia Santiago (Chapter 21), an international recording artist and singer who both embraces new technologies and builds strong connections with fans whether on-stage or online. She takes a balanced approach to AI.

"We cannot deny that we are indeed in a tech revolution! Position and make the shift, but don't throw the baby out with the bathwater," Santiago says. "I see that as we navigate a huge transition, it is wise to hold on to good old-fashioned things like community, relationships, soft skills and values in how we do life and business."

Micro is the new Macro

How about the conferences that creators and vendors rely on for brand awareness, networking and education? Overnight, the pandemic led to even the biggest industry events going virtual. Now that in-person is an option again, what will the future of events look like and which format — in-person, virtual or hybrid — should you invest in as an organizer, speaker, sponsor or attendee?

Laura Clapp Davidson (Chapter 67) is the head of the marketing development team at Shure and a familiar face at big podcasting and tech conferences. Yet Davidson is expecting a shift in industry priorities.

"I predict we will see a movement towards smaller, more personalized creator conferences and events and a move away from large-scale shows," Davidson says. "People are craving connection, knowledge and skills, and this platform seems to hold up to those needs."

Marisa Cali (Chapter 18) produces events at Be Present LLC and, like Davidson, sees smaller events trending. "Micro-events are golden for business owners trying to stand out," says Cali. "High-touch, immersive experiences that allow you to be present for those intentional conversations and serendipitous moments will continue to happen."

When it comes to micro-events, Cali isn't just talking about the size of the crowd; she's also referring to the time commitment asked of attendees. "Micro-events, especially those that are online (virtual) like webinars and video podcasts, don't have to be all day either," Cali says.

No stranger to big events, Brian Wallace (Chapter 61) is a fixture at SXSW as a speaker and infographic marketing agency founder. Wallace also envisions the demand for more small-to-midsize events in the coming year. As a result, he's organizing his own regional event — The Innovate Summit — in May 2024 in Owensboro, Kentucky.

"We will see a rise in more thoughtful, focused and targeted regional, in-person events taking shape in the coming years," Wallace says. "People crave human contact in person, especially given the loneliness many of us endured through Covid."

This shift to smaller and more personal connectivity is mirrored in the fabric of community — a vital lifeline in the quest for audience growth. The traditional bastions of organic reach and platform notifications give way to intimate spaces where creators can forge genuine connections. Email marketing emerges as a linchpin, as home studio expert Junaid Ahmed (Chapter 16) suggests, while corporate YouTube lead Nicole Sanchez (Chapter 5) and Phil Gerbyshak (Chapter 47), a sales and leadership expert, emphasize the power of interactive livestreaming and the potency of network participation in amplifying one's message.

In addition to smaller events, brands are also putting more money into smaller creators and perhaps moving away from large deals with mega-influencers. Digital marketer Julia Jornsay-Silverberg (Chapter 91) says, "There is so much power that small creators have in building relationships with, and therefore influencing, their audience. I am excited to see more small creators get opportunities to work with bigger brands."

That's not all that could go small in 2024. Expect micro-content to find its way onto *American Television*. "Episodic story-arc, SHORT-FORM storytelling will really take off in 2024 (thank you, TikTok, Reels, Shorts and our digitally-rewired brains) and further migrate to traditional linear TV and streaming channels," says Jon Burk (Chapter 17), marketing lead at American Dreamer Media.

"Brands are feeling the appeal of this short-attention-span storytelling as a way to reach, tether and resonate with Gen Z (and younger)," Burk says.

If you spend any time on LinkedIn, you can probably attest to the failure of massive spray-and-pray campaigns when checking your inbox. At best, they bring indifference to the offers. At worst, they damage the credibility of the senders.

LinkedIn expert Karen Yankovich (Chapter 64) advises her clients to take the less is more approach. "Outreach to 5-10 people each week, micro-targeted and researched, with the intention of having an actual conversation, Yankovich says. "That's where business is happening, and I see that only being stronger in 2024 and beyond."

Smaller also is likely to gain popularity in the online learning space. Educator and business owner Katie Hornor (Chapter 70) sees smaller training segments as fitting best with current lifestyles. "More people will pay for online programs that are pre-recorded in smaller 5-10-minute chunks (over live interactive trainings or long webinars) as our society once again prioritizes time around in-person learning and activities," says Hornor.

Creator Self-Care

Hustle culture was all the rage in the mid-to-late 2010s, but you don't hear it talked about as much anymore. In fact, more creators have been speaking out about the negative side of a life lived largely online and always on deadline. The pressure to create, to always come up with new ideas and to be on platforms to respond to comments and engage in other people's content is exhausting — even while just imagining it as I type this.

While a case can be made that some forms of podcasting, livestreaming and video creation benefit from a volume strategy, the ROI of an always on social media lifestyle and massive repurposing campaigns is becoming more questionable. Not only is organic reach largely dead on most social media platforms, but many of the posts are scheduled and responses automated. The opportunities to connect and grow are still there, but they are harder to reach and take more effort than they did even five short years ago.

LinkedIn influencer Brian Schulman (Chapter 20) and functional medicine practitioner Claudine Francois Chapter 13) spotlight the burgeoning dialogue around health and well-being in digital media, while former livestreamer and *Confidence on Camera* author Lottie Hearn (Chapter 51) advocates for a balanced life that harmonizes online engagement — or replaces it — with offline rejuvenation.

Thus, creators need to focus on what they do best, what their audiences respond to and what they enjoy. If you can combine at least two of those three elements, you have something worth spending time on. Taking breaks, prioritizing your physical and mental health and ensuring time is spent with loved ones is an essential part of a sustainable career in content creation and online business.

In the best of circumstances, AI can help you do the productivity tasks faster and better while enabling us to focus our attention on that which makes the difference — hosting our shows, creating our content and engaging with our audiences, leaving us time to take breaks and refresh for the next round.

It's Crowded and Quiet

Meanwhile, there's more content and competition for attention in digital spaces than ever, yet creators sometimes wonder where did everyone go? Do people still get notified when you go live or upload a new video? Platforms are prioritizing different types of audio and video content than they did just a few years ago. Audience tastes are changing, especially when it comes to the younger generations.

And don't expect much help from most social platforms in putting your content in front of people as you attempt to grow your audience. The era of wide organic reach is over, but people are still building audiences and monetizing content. Our contributors share how they plan to keep the momentum going in 2024, and how you can too.

Does that mean going all-in on Shorts, Reels and TikTok is best for growing your audience?

Or is it all about long-form podcasts like so many new media celebs are doing? And do you need video for your podcast? (And is a video podcast on YouTube really a podcast? Does that even matter or is it semantics?)

Plus, what do we need to know about *Podcasting 2.0*?

Hall of Fame podcaster Rob Greenlee's (Chapter 2) insights into the convergence of audio and video mediums highlight a future that transcends traditional definitions. This volume embraces a broad perspective on content creation, recognizing that the winners in this space will be those who can seamlessly blend various mediums and technologies while never losing sight of the primacy of connecting with the people who show up for you.

It's All About YOU

If I asked someone from your audience why they care about your podcast, I doubt anyone would answer, "the show notes." For video creators, would anyone say they go to your videos week after week just for the description?

We need to find ways to assign these tasks to AI tools, posthaste. And be thrilled about that.

Thumbnails, repurposed clips for social media, promotional posts, quote graphics — all are elements that can be important parts of growing awareness for your content and keeping your audience informed, but they aren't the key ingredient that ultimately determines if you sustain that audience.

That indispensable ingredient is the one thing AI can't replace, at least not yet — and that is YOU. You, the show host. Your unique ideas, experiences and personality. Your ability to tell a story. Your personal and professional adventures that take the general and generic and add specifics — real-life examples and situations that put meat on the bones of your content.

The one thing AI can't do as well as you is host your show. It's fine if it does show notes in seconds that aren't quite as good as the ones you spend an hour on. It's wonderful if it does them even better and still saves you 59+ minutes. In any cost/benefit analysis, that variance in the quality of your show notes hardly moves the needle on your bottom line. Livestreamers, podcasters and YouTubers aren't competing on show notes. Your primary differentiator — the only one that truly matters — is the YOU in your livestream video, podcast recording or YouTube upload.

So, this isn't a book about AI; it's a book about how AI can make you faster and more efficient at things that matter but aren't the main ingredient that separates you from the next creator — which makes the fundamentals of content creation more important than ever.

And yet the potential for AI to make real-time improvements in our core content is coming, and not in ways that displace the creator. Larry Roberts (Chapter 6), a frequent speaker and workshop leader educating audiences on the power of AI, says, "Creators will be able to leverage AI tools for real-time video editing, background noise reduction and even real-time language translation, making livestreams more professional and increasing overall accessibility."

Roberts adds, "Livestreaming will become more interactive, with AI generating real-time audience engagement features like polls, quizzes and Q&A sessions. AI will be able to analyze viewer feedback during the stream, allowing creators to adjust content on the fly for better audience engagement."

AI could even help with monetization. "Finally, through advanced data analytics, AI will help creators and brands identify their most effective monetization strategies," Roberts says. "This includes optimized ad placements, sponsorship recommendations and personalized merch."

We welcome these AI additions to our content creation process as improvements, but the central truth is that once all strategies and tactics, tools and tricks are implemented, the game still comes down to the players on the field. And the most important player is YOU, the creator.

In these pages, you will find not only predictions but pathways — avenues to navigate the complex web of digital media with AI as your ally. You will discover strategies to leverage AI for efficiency without compromising the authenticity that defines your work. You will learn how to adapt, how to thrive and how to remain true to yourself in a world that is constantly evolving.

Welcome to a journey through the converging paths of AI and human ingenuity, where the insights of seasoned experts arm you with the knowledge to thrive in the digital media frontier of 2024 and beyond. This is not just a book about the future; it is a roadmap for creating it.

Ross Brand
#100Predictions

P.S. I've added a glossary for the first time as well as updated my creator toolkit with new gear and software recommendations.

PART II
PREDICTIONS

1. Chetachi A. Egwu, Ph.D.

"We'll see a rise in companies improving the look and effectiveness of AI-generated video, providing opportunities for seasoned and novice video makers to use AI beyond generating video clips. "

CHETACHI A. EGWU, PH.D.
IndieSoup Media

The events of 2023 will cause the independent content creator and social influencer to become more important for video and general media creation in 2024!

The Hollywood Grasp on Content Decreases

Hollywood and the traditional media industries as we know them are essentially over. The two concurrent strikes, Writers Guild of America and SAG-AFTRA, may have finally solidified that the future of entertainment does not solely rest in the hands of broadcast, cable or even the big streamers.

This will lead to a continued rise of independent content and the consideration of social influencers as a true source of content rather than just as marketing partners. Social media will also see more scripted content.

The Move Toward AI-Produced Video

Without a doubt, the biggest mass trend of the 2023 was AI, from art to writing and beyond. However, realistic looking and effective video has been a challenge.

In 2024, we will see a rise in companies improving the look and effectiveness of AI-generated video. This will provide opportunities for seasoned and novice video makers to use AI beyond generating video clips.

The Growth of Social News

Traditional news programs have a problem — they are not attractive to the next generation of potential news watchers. In fact, many employ the social media strategy of simply repurposing linear content on their social channels. Coupled with the continued salience of social media for younger Millennial, Gen Z and eventually Gen Alpha audiences, this potentially gives rise to "social news."

We have witnessed that younger audiences tend to get their information from social media. News outlets like Newsy and Cheddar that are aimed at the younger Millennial and Gen Z audiences have demonstrated that it is possible to get younger audiences interested in news.

Even traditional television is taking note. In April 2023, John Oliver actually made an episode of his show, Last Week Tonight, for Gen Z. As such, content creators may increasingly begin to replicate the Cheddar/Newsy style for short form news content in places like TikTok and Instagram.

Traditional Industries Go Beyond Influencer Marketing and Reality Shows

The Hollywood strikes may have had the unintended effect of encouraging traditional media platforms to look elsewhere for content and talent, as was the case with the 2007 WGA strike and the rise of reality programming. All one needs to do is look at the earning power of the top 50 content creators on the 2023 Forbes list to know that they hold peoples' attention. Some on the list have already obtained more traditional deals for linear media.

I see more of this happening in 2024, as the traditional media industries realize the power of these individuals to draw eyeballs beyond influencer marketing and reality programs.

Chetachi A. Egwu, Ph.D., aka Dr. Tachi, is a media professor, filmmaker, content creator and journalist with a passion for creating and educating in the ever-evolving world of media. She is the founder of IndieSoup Media, a content network for independent creators, the host of the weekly livestream MediaScope and co-host of the TV Channelling and Pop Unfiltered podcasts.

2. ROB GREENLEE

"Improvements in content creation and publishing tools simplify all the time-consuming processes, allowing creators to focus more on content."

ROB GREENLEE
Spoken Life Media

Navigating the Near Future of Audio and Video Podcasting

In the rapidly evolving digital landscape of podcasting and livestreaming, significant change trends are shaping the future of content creation, distribution and monetization. These changes highlight the consolidation of content providers, leading to larger audiences for fewer large-scale audience-reaching creators.

This trend requires creators to innovate and refine content to maintain listener interest. Creators are diversifying monetization methods, combining audio and video to maximize reach and revenue through subscription models and dynamic ad insertions.

The rise of *Podcasting 2.0* innovations is also noteworthy, introducing features that enhance listener engagement, Bitcoin-related sharing of *Satoshi* coin value relationships like has never been possible before, and giving creators more control and analytics.

Advanced noise reduction AI technology, like the *insoundz* and *Nomono* platforms, transforms podcast production, making high-quality content creation more accessible and efficient, especially for creators at all levels, including independent podcasters.

Additionally, improvements in content creation and publishing tools simplify all the time-consuming processes, allowing creators to focus more on content.

Despite the saturated market, there is still an opportunity for new and independent creators, particularly as larger media companies scale back due to advertising spending pullbacks. Successful podcasters will consistently produce high-quality content as the overall podcast audience grows. The podcasting and livestreaming industry tools are maturing and becoming more sophisticated.

Creators who adapt to these changes, leverage new technologies and focus on quality and engagement are poised to succeed in this dynamic market. The coming year will offer significant opportunities for those ready to embrace these evolving trends.

Rob Greenlee, a 2017 Podcast Hall of Fame inductee, has over 20 years of experience, starting in radio in 1999 and pioneering podcasting since 2004. He's held VP-level executive roles at industry giants like Xbox, PodcastOne, Libsyn and Spreaker, and is renowned for his profound impact and expertise in the evolving podcast landscape.

3. CLARICE LIN

"In the digital marketing landscape of 2024, the foundation of success for e-commerce businesses lies in effective branding."

CLARICE LIN
BaselineLabs

In the digital marketing landscape of 2024, the foundation of success for e-commerce businesses lies in effective branding.

Only when a brand has established a strong and unique identity can paid marketing strategies, such as Google Ads, be successful in expanding the customer base and, more importantly, in retaining those customers as a consistent source of revenue.

It's vital to recognize that while AI can assist in generating marketing assets and optimizing campaigns, the core uniqueness of a brand relies on the creativity that stems from the human brain and is not part of AI's training algorithm. Brands should continue to harness the creative power of human minds to craft distinctive and compelling brand narratives.

AI should be viewed as a supportive tool to enhance marketing efforts. It can help streamline processes, improve targeting and analyze data, but it should not replace the creative ingenuity that sets a brand apart. Brands must double down on their uniqueness, leveraging the creativity and innovation of their human teams to craft authentic and resonant brand identities.

Simultaneously, marketing automation should be employed to streamline routine tasks, freeing up time and resources for the creative aspects of branding. In this highly competitive landscape, brands that strike the right balance between human creativity and AI support will have a stronger chance of standing out and retaining customer loyalty. Simply relying on automation without a unique human touch risks the brand getting lost in the sea of competitors.

The digital era has brought unprecedented opportunities for e-commerce businesses to thrive, but it has also intensified the competition. Building a strong brand is not only a way to stand out in the crowded marketplace, but it's also a means of building trust, forging connections and retaining customers.

Your brand is your promise to your audience, and it's the essence of what sets you apart in the world of e-commerce.

Clarice Lin is the founder of BaselineLabs, an e-commerce agency that helps Shopify store owners to raise their revenue by 250%-350% through bespoke Google ads strategies. Her YouTube channel has over 3,000 subscribers to help online stores become sophisticated with their marketing strategies. Prior to founding BaselineLabs, she spent over 10 years at BP, Microsoft and Haymarket.

4. Mitch Jackson

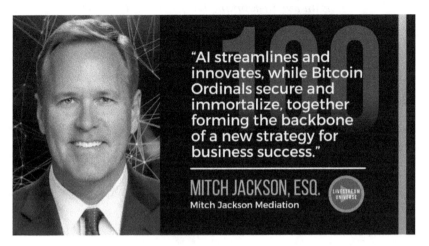

"AI streamlines and innovates, while Bitcoin Ordinals secure and immortalize, together forming the backbone of a new strategy for business success."

MITCH JACKSON, ESQ.
Mitch Jackson Mediation

Revolutionize Your Strategy with AI and the Security of Bitcoin Ordinals

In the 2024 digital landscape, two innovations stand out to me: Artificial intelligence and *Bitcoin Ordinals*.

AI is the key differentiator for businesses and content creators, offering unparalleled efficiency and customer engagement. It's not about keeping pace; it's about setting a vision that leaps ahead.

Meanwhile, Bitcoin Ordinals carve out a new path for digital ownership, providing a secure, history-embedded alternative to smart contracts, all under the security of the *Bitcoin blockchain*. This isn't just technological adoption; it's about creating a lasting digital legacy for your products and services.

Embracing these technologies is essential for leadership in the new business world. AI streamlines and innovates, while Bitcoin

Ordinals secure and immortalize, together forming the backbone of a new strategy for business success.

Mitch Jackson is an award-winning lawyer with 30+ years of experience. As a 2013 California Litigation Lawyer of the Year, he's adept at transforming conflict into cooperation, helping clients with traditional civil disputes and also new AI, Web3 and metaverse challenges.

5. NICOLE SANCHEZ

"Short-form content will continue to explode; however, monetization and community building will still be more robust in the long-form category."

NICOLE SANCHEZ
Snowflake

In 2024, the use of video in marketing will continue to skyrocket, dominating the digital landscape.

Demand for video content for virtual and in-person events keeps expanding and brands are doubling down on livestreaming as that connection with the audience in real-time fosters authentic interactions and a sense of community and trust.

Companies will create more tailored video content that speaks directly to their target audience. This customization will enhance customer engagement and foster stronger brand loyalty.

It's all about AI! Does one person in a thumbnail look grumpy? Use AI to make that expression more pleasant. Not sure which thumbnail to use? There are tools to A/B test them. We'll also see more AI-powered recommendation systems that analyze viewers' preferences and behavior to suggest relevant content, helping creators reach their target audience more effectively and streamline content creation.

Short-form content will continue to explode; however, monetization and community building will still be more robust in the long-form category.

Content creators/influencers will continue to be a force to be reckoned with by brands. These creators forge solid connections and rapport with their audiences, and followers look to them for recommendations, advice and opinions. This relationship makes influencers powerful advocates for brands and products.

Also, content creators create engaging and compelling content that resonates with their specific niche or target audience. This content creation prowess enables them to capture attention and drive conversations around brands or topics, amplifying brand awareness and visibility.

Furthermore, influencers deeply understand their audience's preferences and interests, allowing them to deliver tailored messaging that resonates effectively.

Nicole Sanchez is an executive producer and YouTube audience development strategist. She leads the video team at Snowflake and runs Snowflake's three YouTube channels. Nicole also has her own YouTube channel "The Nicole Sanchez," a luxury hair care and lifestyle channel.

6. LARRY ROBERTS

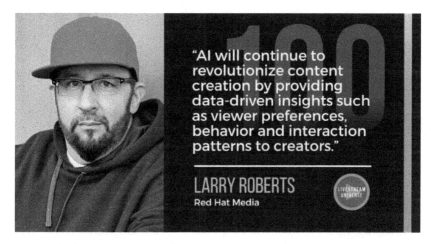

"AI will continue to revolutionize content creation by providing data-driven insights such as viewer preferences, behavior and interaction patterns to creators."

LARRY ROBERTS
Red Hat Media

In 2024, the integration of AI in livestreaming and digital media will not only streamline content creation and enhance viewer experiences, but it could also pose new challenges.

Creators and brands will need to maintain a state of continual learning to stay informed about the latest AI advancements. As AI becomes a more integral part of the digital media landscape, adopting a strategic and responsible approach will be the key to success.

AI will also continue to revolutionize content creation by providing date-driven insights such as viewers' preferences, behavior and interaction patterns to creators. This will lead to highly focused content recommendations and targeted ads, which will significantly improve audience engagement and support.

Creators will be able to leverage AI tools for real-time video editing, background noise reduction and even real-time language translation, making livestreams more professional and increasing overall accessibility.

Livestreaming will become more interactive, with AI generating real-time audience engagement features like polls, quizzes and Q&A sessions. AI will be able to analyze viewer feedback during the stream, allowing creators to adjust content on the fly for better audience engagement.

Finally, through advanced data analytics, AI will help creators and brands identify their most effective monetization strategies. This includes optimized ad placements, sponsorship recommendations and personalized merch.

As the founder of Red Hat Media and co-host of the Branded podcast, Larry Roberts is aptly seen as a thought leader in the content creation and AI space. Regularly teaching workshops and hosting speaking engagements to help creators utilize new technology, he continues to push business owners out of their comfort zone and into a chapter of innovation and growth.

7. JODI KRANGLE

"Sensory branding will become more widely used. Tying in a brand experience with not just sight and sound but also a scent or a taste or a touch."

JODI KRANGLE
Piece of Cake Voiceovers

My predictions are all about the increasing use of audio in advertising and our daily lives — and how big an impact it'll make in the coming years.

The use of sound to connect with your clients is becoming more of a necessity than a luxury as the years pass. The pandemic solidified that necessity. Our isolation made us look for increasingly meaningful ways to connect remotely. Sound has been a huge part of that.

One of the biggest examples of this is the rise of podcasting and how mainstream it's become. YouTube is now one of the premier places for podcasts to be discovered, but generally that's only for podcasts that also have a video component (rather than those that are merely static pictures with audio — the way that a lot of podcast hosts automatically create for their subscribers).

Even so, while video is important for discoverability, without good sound, it won't hold the attention of an audience for long. And that's more about not putting undue stress on a listener's ears as opposed to having "perfect" sound.

We have our attention split in so many different directions daily, but sound is something we can experience while we're doing other things. So, it fits uniquely into all the spaces we have in our lives.

Additionally, audio or sonic branding is becoming more and more mainstream. It used to be something only the larger companies could afford to do, but now even the smaller ones have heard of it and are getting involved in it.

I also predict that in the years to come, sensory branding will become more widely used. Tying in a brand experience with not just sight and sound but also a scent or a taste or a touch. The more well-rounded these experiences can be, the deeper of an impression they'll make — and the more memorable they'll become.

The rise of *Dolby Atmos* will revolutionize the delivery of audio experiences. When media companies are able to take full advantage of it, we'll truly be in a golden age for sound.

Jodi Krangle has been a full-time voice actor since 2007 and has worked with clients from major brands all over the world including Dell, Lindt, Pfizer and Kraft. Over the years, she's learned a lot about sound and how it influences people. With over 200 episodes, her podcast on this subject is called Audio Branding: The hidden gem of marketing. Learn more at audiobrandingpodcast.com.

8. Mike Allton

"Businesses will be forced, financially, to re-invest in virtual events, both from a marketing DemandGen perspective, and from a consumption perspective."

MIKE ALLTON
The Social Media Hat

During the pandemic, for obvious reasons, virtual events of all sizes and shapes exploded. Technology solutions saw explosive growth, and many poured that interest into development and enhancements to improve the event and attendee experience.

But then the pandemic ended and an insatiable hunger for travel and in-person contact took over. Virtual event attendance and utilization plummeted, and many solution vendors were irreparably damaged. In 2024, the pendulum will swing back the other way.

Today, businesses continue to feel the pressures of uncertain economies and global conflicts. While consumer spending in key segments remains high, consumer confidence in the economy is remarkably low, and rumors of a recession persist. As a result, budgets and marketing investments continue to be pinched at the time costs involved with travel continue to rise.

In 2024, businesses will be forced, financially, to re-invest in virtual events, both from a marketing *DemandGen* perspective, and from a consumption perspective. While marketing teams love and appreciate hosting and exhibiting at in-person events like trade shows and conferences, most departments will lack the budget to realistically tackle such events more than once or twice a year.

Yet their sales teams are begging for increased pipeline, so those teams will have no choice but to turn to virtual events which are far more cost effective and easier to stand up more frequently. Your company might be able to host quarterly virtual summits and weekly webinars without significant expense.

At the same time, demand for more virtual events will also increase — the consumption perspective — for similar reasons. Given the choice, executive and financial teams would rather have their teams attend events virtually instead of in person and save on travel expenses.

My advice for all businesses is to be aware of this trend and to be careful to approach it in moderation. People require in-person engagements for both personal and professional reasons — namely, relationship building. So, businesses would be wise to reduce, but not eliminate, in-person elements in their events and training opportunities. This will keep the pendulum from swinging too far in the other direction and may stave off future over-corrections.

Mike Allton is an international keynote speaker, an award-winning consultant and author at The Social Media Hat, and head of strategic partnerships at Agorapulse, where he strengthens relationships with social media educators, influencers and partner brands.

He has spent over a decade in digital marketing and brings an unparalleled level of experience and excitement to the fore.

9. JUDI FOX

"Memberships and course sales success will rely on businesses correctly choosing what can be automated and what parts of client success need to be kept high touch."

JUDI FOX
LinkedIn Business Accelerator

Welcome to 2024! As AI improves creator processes like making podcast editing faster, I am predicting that AI will reshape how creators gain ideal client trust.

Here are several #FoxRocks predictions:

- Algorithms will become better due to AI tools, which means creators who deliver unique, quality content will benefit.
- Memberships and course sales success will rely on businesses correctly choosing what parts of their business can be automated and what parts of client success need to be kept high touch.
- Referrals and recommendations based on trust will become more important for event attendance and online business.

We rely on social signals like post shares, likes and comments to tell us something is successful for us to buy or join. But with AI able to quickly do those social proof tasks, we will have to rely on content and conversations online that feels more "word of mouth" to trust we are consuming the "real deal."

2024 is going to be an incredible year for anyone that stays steady as platforms get flooded by overly AI-generated content. The "real deal" creators and businesses will shine through in 2024.

P.S. "Why did AI start going to therapy?"

"It had too many unresolved issues." :)

Judi Fox (#FoxRocks) is a LinkedIn Top Voice in Sales and Marketing. She is listed as a Top 10 LinkedIn Coach in Yahoo Finance and a featured speaker at Social Media Marketing World, VidSummit and more. With 18 years of experience in business, Judi founded the LinkedIn Business Accelerator Method. Clients implement this LinkedIn method to achieve more business and over 1 million content views in 90 days.

10. DALE L. ROBERTS

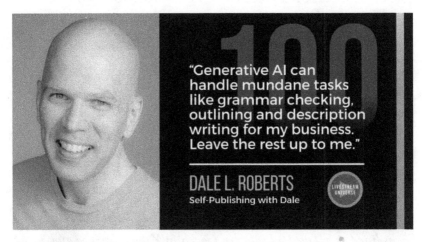

"Generative AI can handle mundane tasks like grammar checking, outlining and description writing for my business. Leave the rest up to me."

DALE L. ROBERTS
Self-Publishing with Dale

I don't think it takes a genius to point out that generative artificial intelligence (AI) has taken the world by storm over the past couple of years — for better or for worse. Creators can use AI to generate videos, text and images all through a few prompts.

However, the hype beasts of AI focus entirely too much on what generative AI can be rather than what it is now. In its current iteration, AI needs human intervention to create passable content for consumers.

Savvy online entrepreneurs understand it requires more than a single prompt to write a novel or produce a movie. Even creating passable AI-generated videos is a bit of a stretch. Just look up AI-generated animations and you'll have a fresh resource for nightmare fuel.

Generative AI isn't all bad, just as it isn't all good. How online entrepreneurs perceive AI will make a difference in how efficient they can be in their future business.

In the next few years, AI is only going to improve. Will we eventually see single prompt options with superior AI-generated outputs? Absolutely! Will everyone be on-board for that? Possibly not.

For instance, I'm an author and video creator. A large part of why I got into this business was to exercise my creative muscles. I can't imagine running my business without using some of my creativity.

Leveraging generative AI to create all my content doesn't quite align with my goals. Generative AI can handle mundane tasks like grammar checking, outlining and description writing for my business. Leave the rest of it up to me.

Using AI helps optimize my workflow, so I have more time to channel my creative energy into other aspects that fulfill me. Anyone leveraging AI works at a greater advantage over folks who refuse to embrace the inevitable future. Rather than fight it, they figure out how it best suits their business model to gain an edge on their competition.

For anyone opposed to AI, they'll have to continue working at a snail's pace in comparison to those who embrace artificial intelligence in a safe and ethical manner. Will you give generative AI a shot? If not, what's holding you back?

Simply crossing your arms and stomping your feet won't put the toothpaste back into the tube. Artificial intelligence is the cat that's been let out of the bag. Either run with it or be left in its wake.

Dale L. Roberts is an award-winning author and video content creator. He's published over 50 books, interviewed over 150 guest experts in publishing and produced over 1,000 videos about the business of writing and self-publishing since 2014. When he's not working, he enjoys reading, working out and traveling.

11. VALERIE MORRIS

"We're going to continue to see a huge shift to AI streamlining and infiltrating almost anything we do."

VALERIE MORRIS
Tintero Creative

We're going to continue to see a huge shift to AI streamlining and infiltrating almost anything we do. That's only going to make community and authentic humanity more valuable.

When people know they are having an authentic experience, it has that much more weight because people are wanting more and more of that. Yes, AI has a huge opportunity in marketing and communications, but there's something special about real-time, authentic interaction and insight.

Find ways to make your own content authentic and leverage productivity tools and AI to help you do your authentic content faster and more efficiently. This is where podcasting and live video have a prominent role. These are ways you can lean into authentic content, showcase your personality, build trust and build your brand.

This is also where mistakes and imperfectly excellent production can help you out too. One of the best signs of authenticity is imperfection. You obviously want to put out quality content but embrace any of the blemishes because they actually help give you more credibility as an authentic voice.

Valerie Morris is a social media and content marketing strategist. She founded Tintero Creative, a digital agency based in Colorado. She also runs programs to help authors launch books. Valerie is a speaker, trainer, podcaster and bestselling author. When not online, she can be found homesteading or looking for the best queso in town.

12. ROGER WAKEFIELD

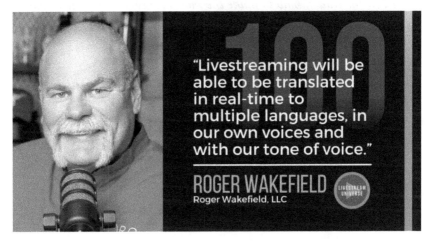

"Livestreaming will be able to be translated in real-time to multiple languages, in our own voices and with our tone of voice."

ROGER WAKEFIELD
Roger Wakefield, LLC

In 2024, I predict that livestreaming will be able to be translated in real time to multiple languages. In our own voices and with our tone of voice.

Not only translated, but AI will be able to modify our video broadcast in a way that our lips will look like we are speaking that language. What this will do is allow us to communicate live with people from around the world in their language while commenting over a video livestream.

Imagine being able to communicate in your native language through a scheduled livestream, read live comments and give amazing feedback — while on the other side of the world, someone is listening in their native language and commenting, talking to you, in real time.

At this point our brand is **global**!

Roger Wakefield is the voice of the trades. Showing people how to get started in the trades, how to make the most money in the trades, how to start their own businesses, and how to G.R.O.W. and make their phone ring!

13. Claudine Francois

"We'll see an explosion in consumer-driven content in the following three categories of health and wellness: Wearables, personalized medicine, and health communities."

CLAUDINE FRANCOIS
In Good Clean Taste

The health and wellness industry has grown 5-10% each year and is showing no signs of stopping. And it's no longer all about CrossFit and juice cleanses. It's getting personal.

Many of today's wellness consumers are getting tired of the traditional, vague advice to "eat more vegetables and workout five times a week." Instead, they are looking to figure out their personal metrics:

- What foods (and in what order) keep my blood sugar steady?
- How much sleep is ideal for MY body (and at what times of the month or after which specific activities)?
- What supplements does my body need — right now — for peak productivity and performance?

These are the questions that are popping up in the health and wellness podcast scene.

So, what is my prediction for 2024? That we will see an explosion in consumer-driven content on the major social and digital platforms in the following three categories of health and wellness:

Wearables — Getting Connected

Yes, everyone's got a smart watch but how many are wearing a CGM (Continuous Glucose Monitor) to track their glucose spikes so they can prevent weight gain, brain fog, and diabetes? And what about the little ring some wear on their finger to track their sleep?

Getting a good night's sleep is step one in getting a head-start on the day (sleep-deprived people have more stress and eat more calories, on average) and getting your gold star from this device could be just the thing you need to make it "worth your while" to get those zzz's.

From menstrual cycle trackers to heart rate monitors, pretty much anyone with a platform on YouTube, livestreaming or podcasting is talking about the latest wearable tech — because we love data!

Personalized Medicine

And speaking of data, the days of waiting an hour for the doctor to spend five minutes flipping through your chart just to tell you to "eat your broccoli and do some more cardio" are over! Consumers are demanding to know much more about their health — and not just when they are "diagnosable." The market for pre-disease testing is HUGE. And podcasters and livestreamers are bringing in the experts.

Want to avoid getting that diagnosis everyone in your family has? From functional medicine to specialty full-body scans (yes, a lot like in that Jodie Foster movie!), the world of personalized medicine is moving at record speed and the influencers in this space are taking note.

Health Communities

If you've watched the "Blue Zones" documentary or read any of the books on habit building, you'll know that people who surround themselves with friends who have similar health goals tend to stick with those goals. Podcasters, livestreamers and other influencers are building those communities within their platforms to engage their audiences and create a social network of like-minded individuals: People who want to improve their health outcomes by taking back their power.

So, whether you want to know your personal biomarkers, hack your way to healthy longevity or connect with collaborators in your health journey, 2024 is the year I predict we will have an exponentially-growing resource of influencers bringing us the information we need to jumpstart our health — and thrive.

*Claudine Francois, **a Board-Certified Functional Medicine Practitioner, empowers women to reclaim their health narratives and find lasting solutions. With expertise in uncovering the root causes of chronic issues, she uses sophisticated tests and a holistic approach to craft personalized protocols for energy, clarity and vitality.***

14. NICK NIMMIN

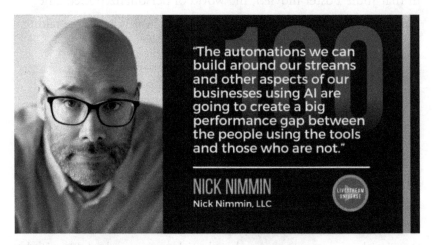

"The automations we can build around our streams and other aspects of our businesses using AI are going to create a big performance gap between the people using the tools and those who are not."

NICK NIMMIN
Nick Nimmin, LLC

Over the last handful of years, we've seen the continued rise of livestreaming due to low barrier to entry into streaming from companies like StreamYard and the popularity of livestreams across all of the major social platforms. It's going to keep getting easier and more popular.

However, AI is already changing what's possible for streamers. From being able to use AI voice to make brand assets, ChatGPT for ideas and run of show outlines or giving you questions to ask your guests, TubeBuddy's AI that predicts the click-ability of your thumbnails, to easily repurposing your stream content into vertical content with just a few clicks using tools like OpusClip.

Everything is changing and by this time next year, AI will make streaming easier and a lot more fun! In addition, the automations we'll be able to build around our streams and other aspects of our businesses using AI are going to create a big performance gap between the people using the tools and the people who are not. The future of livestreaming is looking brighter than ever!

Nick Nimmin is a YouTube content creator and specialist helping creators and brands boost their online presence through YouTube.

15. JUNAID AHMED

"As the lines between podcasting and video blur, one thing is clear: The future is visual, interactive and deeply personalized."

JUNAID AHMED
Home Studio Mastery

The Rebirth of Video Podcasting

Digital media predictions for 2024:

1. The Rise of Video Podcasting on YouTube: With Google's decision to fold its podcast division, redirecting its audience to YouTube Music, there's a clear indicator: The future of podcasting is visual. This will lead to an exponential rise in video podcasts, marrying the depth of podcast content with the visual allure of video. Podcasters will capitalize on YouTube's vast audience, blending rich audio content with compelling visuals, creating immersive experiences that pure audio formats can't match.

2. Discoverability and Engagement: On YouTube, podcasters will enjoy enhanced discoverability thanks to the platform's robust AI-driven recommendations. However, to truly stand out, they'll need to focus on thumbnails, metadata and episode descriptions. This new ecosystem will also spur the creation of tools and services tailored to optimize podcast content on YouTube. Keep in mind that you are using YouTube as a platform to not only expand the reach of your podcast but also to build out your brand through your branded studio setup.

3. The Importance of Quality Production: As video podcasting becomes more prevalent, the focus on production quality intensifies. High-quality video and audio production are becoming essential for standing out in an increasingly crowded market. A survey reveals that podcasts with higher production values retain 30% more listeners than those with basic or poor production.

4. Investment in Equipment and Skills: There's a growing trend of podcasters investing in professional-grade equipment and enhancing their production skills; this includes better cameras, lighting, sound equipment and software for editing and effects.

5. Email Marketing — The Unsung Hero: As creators juggle between audio and video formats, email marketing will serve as the linchpin. Newsletters will offer subscribers curated content, from episode summaries to behind-the-scenes peeks and direct links to the latest uploads. This personalized touch will further deepen the bond between creators and their community.

6. Interactive Features and Real-Time Engagement: YouTube's native functionalities, such as Super Chats and Community Tabs, will be a boon for podcasters. They can host live podcasting sessions, fostering real-time interaction and community-driven content.

7. Personalization and Adaptation: Advanced AI will usher in an era of hyper-personalized email marketing. Expect dynamic newsletters, adapting content based on user interactions, offering tailored podcast and YouTube recommendations.

In 2024, as the lines between podcasting and video blur, one thing is clear: The future is visual, interactive and deeply personalized.

Junaid Ahmed recognizes the monumental shift towards video podcasting. Through Home Studio Mastery, he is on a mission to empower entrepreneurs with the goal of helping creators establish state-of-the-art, versatile video podcast studios. Junaid also hosts the Hacks & Hobbies podcast where he brings forth the Passion to Profit stories of entrepreneurs from all over the world in various industries.

16. JENNYQ

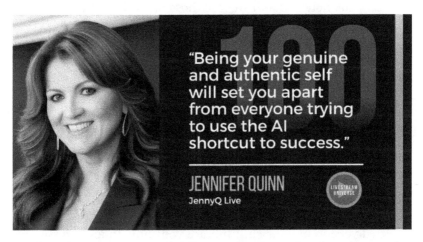

"Being your genuine and authentic self will set you apart from everyone trying to use the AI shortcut to success."

JENNIFER QUINN
JennyQ Live

For the 9th year in a row, I'm honored to have the chance to make a prediction for Ross Brand's Annual Livestream Universe Predictions.

Last year, I highlighted the critical importance of developing a Web3 business strategy, and this time around, I want to talk about the impact of Artificial Intelligence (AI) on customer-facing marketing.

AI can be a double-edged sword. On the one hand, it can supercharge our marketing efforts, but on the other, it can make things feel a bit robotic and distant, disrupting that human connection we value so much.

My "aha" moment came while watching a video ad. People in the comments section suggested that the presenter was just an AI, not a real person. Even though I believed he was an actual human, the excessive use of AI made it look otherwise — the video lacked that personal touch we all appreciate.

The real issue here is that if we lose those natural quirks like the 'ahs,' 'ums,' pauses, and even a bit of laughter when we stumble over a word, we risk coming across as inauthentic. People are very good at spotting when something's not quite right, and if they sense an overreliance on AI, trust goes out the window.

To quote a dear friend of mine, human behavior expert, the late Dave Lakhani, "The quickest way to build trust is to show up the way people expect you to."

I can guarantee that no one expects you to show up as an AI-perfected version of yourself. Embrace your humanness, be you, be real and connect to other humans.

In 2024, being your genuine and authentic self will set you apart from everyone trying to use the AI shortcut to success. So, what we need to do is strike a balance.

We should use AI strategically in our marketing efforts, not let it take over completely. This way, we can keep that genuine, human connection intact, especially for brands and small businesses connecting with their customers in the B2C world through digital marketing.

Jennifer Quinn "JennyQ" has been a digital marketing pioneer on all fronts. As the host of "Facebook Live's First Variety Show" to the launch of the "Being Seen" podcast, her company specializes in serving professionals by using creative and compelling marketing strategies that convert.

17. Jon Burk

"Episodic story-arc, short-form storytelling will retake off in 2024 and further migrate to traditional linear TV and streaming channels."

JON BURK
American Dreamer Media

Over the past decade, we've watched different iterations of video take hold on different platforms and enjoy their 15 minutes of fame.

However, episodic story-arc, SHORT-FORM storytelling will really take off in 2024 (thank you, TikTok, Reels, Shorts and our digitally-rewired brains) and further migrate to traditional linear TV and streaming channels.

In addition, brands are feeling the appeal of this short-attention-span storytelling as a way to reach, tether and resonate with Gen Z (and younger). Case in point is Travel by Design series sponsored by Marriott Bonvoy on Amazon Prime. This is gorgeous content, featuring a first-person narrative over high quality B-roll, punctuated by a cinematic audio mix; warmly personal and relatable.

I expect other brands to follow suit and bring the customer closer by telling high-quality and impactful short stories across digital and linear channels without them feeling overly invasive or too "salesy."

Jon Burk is marketing lead with American Dreamer Media and the Al Roker-produced brand-funded documentary "Gaining Ground: The Fight for Black Land" (slated for its TV debut in 2024). In addition, Jon is CMO/head of development for Analog 77 Films, a media startup based in LA with several film projects in production.

18. MARISA CALI

"Micro-events are golden for those trying to stand out — high-touch, immersive experiences for intentional conversations and serendipitous moments."

MARISA CALI
Be Present LLC

Micro-events are golden for business owners trying to stand out. High-touch, immersive experiences that allow you to be present for those intentional conversations and serendipitous moments will continue to happen.

Previously, you thought that having everyone in one room was the goal. But it's too much and clouds the experience for the attendees. With targeted micro-events, you help attendees feel valued, especially first-timers, and they get more out of the experience.

Micro-events, especially those that are online (virtual) like webinars and video podcasts, don't have to be all day either. Having an in-person event in three months? Why not have a virtual kickoff and multiple touchpoints that continue to make an impact?

It's beneficial to bring mindfulness and accessibility into your event planning conversation early on to maximize an experience. The digital media landscape will thank you with their loyalty.

Marisa Cali has managed and grown her own marketing consulting business since 2015 and embraces the benefits (and challenges!) of the work-from-anywhere lifestyle. Over the last four years, Marisa has had the pleasure of working with exceptional corporate and educational institution partners to execute virtual summits, exclusive fireside chats, inspirational keynotes and video podcasts.

19. JEFF SIEH

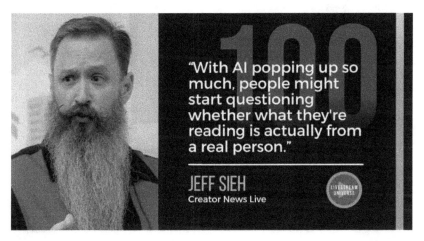

"With AI popping up so much, people might start questioning whether what they're reading is actually from a real person."

JEFF SIEH
Creator News Live

By 2024, I'm expecting to see AI everywhere in apps, kind of like how spell check is all over the place now. But with AI popping up so much, I think people might start questioning whether what they're reading is actually from a real person.

This is great news for livestreamers and podcasters since they bring that real-time, genuine connection that AI won't be able to match.

Jeff Sieh is an international speaker and visual marketing consultant. He hosts the Creator News Live show and podcast and is also "Head Beard" at Manly Pinterest Tips. Jeff has worked with and produced a wide range of content for various companies, including Guy Kawasaki, Kim Garst, Social Media Examiner and Tailwind.

20. BRIAN SCHULMAN

"Expect more digital content related to mental health, stress management and overall well-being, aiming to support and uplift audiences and communities."

BRIAN SCHULMAN
Voice Your Vibe

Heart-centered leadership focuses on empathy, authenticity, innovation, transparency and a genuine concern for the well-being of our communities and others.

In 2024, digital media will see a rise in purpose-driven content that aligns with heart-centered leadership. Brands and organizations will create and promote content that highlights their social, environmental and community responsibilities, solidifying their commitment to making a positive impact on the world.

Digital media will continue to be a powerful tool for driving social impact campaigns and heart-centered leaders and organizations will use their platforms to raise awareness and rally support for important global issues.

Heart-centered leadership stresses the well-being of employees and customers. Expect digital media to feature more content related to mental health, stress management and overall well-being, aiming to support and uplift audiences and communities. Moreover, heart-centered leaders will put accessibility and inclusivity at the forefront, ensuring that content is available to all and actively promotes these values via digital media.

Additionally, look for leaders in digital media (and organizations globally) to focus on fostering emotional intelligence in their teams and content creation. This will lead to more empathetic and emotionally resonant messaging. Brands and influencers will use digital media to share more authentic, personal stories of their struggles and successes, fostering a deeper connection with their audiences — and many will do so in collaboration.

With 1 billion users (Oct 2023), one of the most untapped organic resources, LinkedIn remains a playground of opportunity. While the influencer community is relatively small, for those who are active on the platform, opportunities for brand partnerships will prove themselves to be important, as brands partner with creators to make it easier to cut through the noise and amplify one another's voices.

Another underutilized asset that will become a focus for organizations will be its people. In 2024, organizations will empower their employees to become brand advocates through digital media, encouraging them to share their experiences and perspectives helping to build a more authentic and relatable business brand.

While trends can, and will, evolve rapidly, heart-centered leadership is here to stay and it's important for leaders to be flexible and adapt to changes as they occur.

Heart-centered leadership will serve as a guiding framework for creating content and experiences that resonate with audiences on a deeper and more meaningful level. In the dynamic landscape of digital media, in 2024, it's the heart of it all that matters.

As the founder and CEO, through Voice Your Vibe's ground-breaking masterminds and his heart-centered leadership programs, Brian Schulman has transformed how business is conducted on LinkedIn worldwide for the last two decades. Named 'The King Of Community on LinkedIn' by Forbes, known as the Godfather and Pioneer of LinkedIn Video, and one of the world's premiere livestreaming and video marketing experts, Brian is an 11-time #1 best-selling author and internationally known keynote speaker recognized by LinkedIn as a four-time Top Voice, three-time Top 50 Most Impactful People of LinkedIn, four-time Rising Star and Influencer To Watch and two-time Global Leader of the Year out of one billion, who's expertise, insights and two-time global award-winning LinkedIn LIVE shows of five years and 500+ episodes have been featured on NASDAQ, Forbes, Thrive Global, Yahoo Finance, Bloomberg, Viacom, Roku TV, Amazon Fire, PODTV, The CW, multiple #1 best-selling books, syndicated on smart TV networks and hundreds of shows and podcasts reaching millions worldwide.

21. CLAUDIA SANTIAGO

"With AI quickly changing the efficiency on how we do things, it also gives us freedom of time to create in other ways."

CLAUDIA SANTIAGO
Claudia Santiago

What an honor to be part of this again this year with so many of you! Thank you, Ross!

We cannot deny that we are indeed in a tech revolution. Position and make the shift, but don't throw the baby out with the bathwater.

I see that as we navigate a huge transition, it is wise to hold on to good old-fashioned things like community, relationships, soft skills, values in how we do life and business and ride the wave with current online and social platforms, but it is also key to integrate the new. This is happening quickly, so don't ignore it. Embrace it, educate yourself and implement, implement, implement!

With AI quickly changing the efficiency on how we do things, it also gives us freedom of time to create in other ways. By embracing the new we become more efficient, but it's also an opportunity to allow our "human" qualities to shine.

This is what will set you apart from the AI version of you with the uniqueness of what you bring to the table. Look at the "gold" of the age we are living in, push fear aside and transition well.

Claudia Santiago, a Chilean-born Canadian citizen, is an international recording artist, speaker, writer, producer and entrepreneur.

22. REBECCA GUNTER

"In a world of AI and information overload, the only way around deepfakes and empty promises is to show 'em what's real and raw. Your POV can never be replicated."

REBECCA GUNTER
Stoned Fruit

LIVESTREAM UNIVERSE

Content creators, marketing people and anyone who is nurturing a platform to bring reputations to life, dare not be anemic with your time, effort or quality of work — the Future of Livestreaming & Digital Media is making an impassioned case for long-form content.

If you detest the mediocre and the battle of attrition with noisemakers and na'er-do-wells cluttering the internet and stealing eyeballs from audiences everywhere, do not be discouraged.

In a world of AI and information overload, the only way around deepfakes and empty promises is to show 'em what's real and raw. Your POV can never be replicated.

Rebecca Gunter brings the compassionate collaboration, radical self-belief and immersive support to help you muster the courage to express your value and vibe without inhibition. Founder/Peach-in-Chief at StonedFruit and host of the Stoned Fruit Roll Up, a weekly LIVE, and Business in the Raw, a docu-series following an epic brandventure.

23. Dan Currier

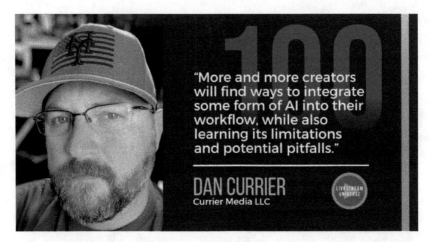

"More and more creators will find ways to integrate some form of AI into their workflow, while also learning its limitations and potential pitfalls."

DAN CURRIER
Currier Media LLC

Last year we saw the expansion of AI in the creative space. This year, as we see it becoming more widely adopted, countless benefits are being revealed. These benefits do not come without caution.

Over the next year, more and more creators will find ways to integrate some form of AI into their workflow, while also learning its limitations and potential pitfalls.

With all of these potential pros and cons to consider, the creator community cannot overlook the raw potential AI represents in providing new levels of efficiency to our workflows, that would otherwise be impossible.

If you have been watching AI from the sidelines, it is definitely time to dip your toe in the ocean of possibilities and see how AI can help you achieve your goals. How does a mouse eat an elephant? One bite at a time!

You don't have to learn EVERYTHING about AI all at once. Try something simple like having ChatGPT write your social media profile or video descriptions. Whether you choose to dive headfirst or just dip your toe, this is the year to get started.

Dan Currier is the CEO of Currier Media LLC. Using social media, Dan shares helpful tips and strategies about making money through online video. Dan's YouTube channel has helped millions build their channels and start earning money online. He has also helped many content creators earn thousands of dollars through the Amazon Influencer Program.

24. JEFFREY POWERS

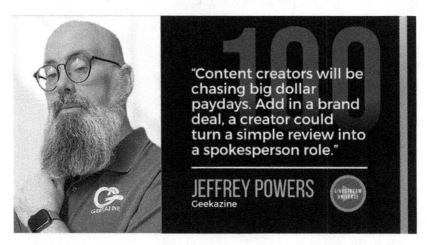

"Content creators will be chasing big dollar paydays. Add in a brand deal, a creator could turn a simple review into a spokesperson role."

JEFFREY POWERS
Geekazine

Shopping features in your media will ramp up in 2024. Amazon, TikTok, YouTube, Walmart and others are already putting carousels on your videos with products you talk about in livestreams, YouTube Shorts (or Instagram Reels) and post-produced videos. The competition for clicks and revenue will increase throughout the services.

Events like Black Friday will be online more than in-store. Starting at 6pm ET, you'll be able to log in, buy, then pick up or have shipped.

Death scrolling and product purchases will start to go hand-in-hand. You won't be able to watch a Short or TikTok without a carousel attached to it. This can bring on health and financial issues, as someone might be scrolling all night and buying beyond their means.

And content creators will be chasing these big dollar paydays. Add in a brand deal, a creator could turn a simple review into a spokesperson role.

Jeffrey Powers **is Geekazine.**

25. LEE UEHARA

"More people are going to continue the trend of monetizing through self-publication with digital distribution services such as Substack and its competitor Beehiiv."

LEE UEHARA
NYC Podcasters.com

1. There will be a surge in using Gmail for sending out newsletters. This feature has been out since last year and is as easy as clicking a button. Yet, it has gone mostly unnoticed for those with Google Workspace accounts.

The functions aren't as plentiful as using a dedicated app. But sending out a newsletter with graphics done from one's Gmail account is fast and doesn't require signing up for yet another program if a subscriber list is very small.

2. More people — with a message and the desire to share it — are going to continue the trend of monetizing through self-publication with digital distribution services such as Substack and its competitor Beehiiv.

As more companies pop up, content creators will also comparison shop for the lowest newsletter platform fees which are based on subscription revenue.

Having said all of that, what's most important is to not only create a message, but to share it. Thus, avoid getting caught up on selecting a method of electronic communication, and Just. Get. Started.

Lee Uehara is a serial podcaster and multi-media content creator and consultant. She is also the co-founder of AAP (Asian American Podcasters Association); the Golden Crane Podcast Awards; and NYC Podcasters.com, an indie collective. Lee is also a rockstar photographer to podcasters and speakers. If you're ever in NYC, hit her up for a coffee-dog walk in Central Park.

26. LIRON SEGEV

"In 2023, we have seen a significant rise in creator YouTube channels getting hacked. This trend will continue to surge into 2024."

LIRON SEGEV
Security For Creators

In 2023 we saw a significant rise in creator YouTube channels getting hacked. This trend will continue to surge into 2024.

Hackers are using AI tools such as ChatGPT to write even more sophisticated emails that are much harder to detect as being "obvious scams." They are making use of *deepfake* videos to convince creators to click on links and are using advanced methods to get around 2-factor authentication.

Based on discussions on the *dark web*, entire creator-hacker rings have creators in their sites. These groups are very aware of how the creator economy works and are exploiting those weak points to take over channels. Something to especially watch out for in 2024 are "brand deals" emails which have become a favorite method for these hackers. The good news is that 93% of hacks could easily be prevented.

In 2024, creators should make security a top priority instead of an afterthought. Hopefully that will minimize the number of channels that find out firsthand how a simple hack decimates their income — most do not recover.

Liron Segev (aka The TechieGuy) is a security tech YouTuber whose channel focuses on educating viewers about the latest hacks and scams so they can keep safe online. Liron is the founder of Security For Creators, a creator-focused cybersecurity firm that helps creators protect their channels and income from hackers.

27. DAVE JACKSON

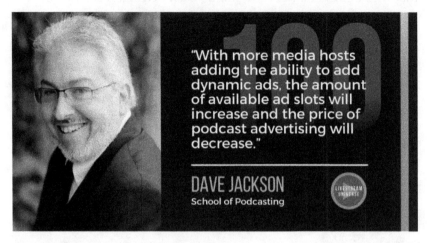

"With more media hosts adding the ability to add dynamic ads, the amount of available ad slots will increase and the price of podcast advertising will decrease."

DAVE JACKSON
School of Podcasting

With more media hosts adding the ability to add dynamic ads, the amount of available ad slots will increase and the price of podcast advertising will decrease. This is not a bold prediction, but more supply and demand.

More companies will emerge to help ensure your podcast is safe for advertisers (for a fee). So, you must pay these companies to prove your podcast is safe while making less money from your ads. This will lead to podcasters using programmatic ads (the ads that pay .003 per download) to cram large amounts of ads into their episodes, making them rival the insane amounts of ads you find on terrestrial radio. This is known as "The race to the bottom."

The most profitable strategy for podcasting will still be promoting your own product and service, and yet new podcasters will continue to say, "I want to start a podcast and get ads." Media hosts that make it easy for podcasters to promote their own services will have an advantage among entrepreneurs.

People with no crowds listening to their podcasts will start up crowdfunding campaigns and become frustrated.

YouTube Music will roll out the ingestion of podcasts via *RSS*. Some will add their show while others will not, leading to the YouTube music app getting less market share than Google Podcasts, which Google closed down.

Dave Jackson was inducted into the Podcasting Hall of Fame in 2018. He started the SchoolofPodcasting.com in 2005. He is the author of the book, "Profit from Your Podcast: Proven Strategies to Turn Listeners into a Livelihood." Over the years he has launched over 30 different podcasts with over 4 million downloads.

28. CAREN GLASSER

"The proliferation of AI tools will increase efficiency and reach, enabling businesses to compete with larger enterprises."

CAREN GLASSER
Caren Glasser LIVE!

LIVESTREAM UNIVERSE

Artificial Intelligence (AI) is becoming an integral part of modern-day life and learning how to harness its power can be exciting and groundbreaking, rather than something to be feared.

The proliferation of AI tools presents a huge benefit for solopreneurs and small businesses as it assists in content creation. It will increase efficiency and reach, enabling businesses to compete with larger enterprises.

Change is inevitable, and it's not uncommon to greet new technology with apprehension. I am embracing AI and integrating it into my workflow in order to work smarter, not harder.

Remember, AI is a tool. It's up to us to use it wisely and beneficially. I believe the future of AI reflects beyond just technology; it's about unlocking human potential and creativity too.

Caren Glasser is the host and producer of multiple online talk shows, including Caren Glasser LIVE!, Life Uncorked, The Author's Spotlight and Minding Your Mental Health. In total, her shows have amassed over 2 million views. In 2018 she was voted one of the 5 top online hosts in the world.

29. Brenden Mulligan

"With YouTube's persistent dominance in content consumption, more podcasters will embrace video, creating dual-format content to cater to both audio and visual preferences."

BRENDEN MULLIGAN
Podpage

Looking ahead to 2024, the landscape of digital media, especially in the realm of podcasting and content creation, is poised for significant evolution:

1. Podcasting's Growth Trajectory: The trend of podcasting as a medium for sharing knowledge is expected to not only continue but also to accelerate.

2. Video Integration in Podcasting: With YouTube's persistent dominance in content consumption, more podcasters will likely embrace video, creating dual-format content to cater to both audio and visual preferences. This approach could leverage YouTube's vast reach, providing podcasters with a broader audience and enhanced engagement opportunities.

3. Content Segmentation and Social Media Presence: The trend of dissecting long-form content into digestible "shorts" or snippets will become more prevalent as podcasters look to reach audiences on social media platforms where brief content performs best. This strategy serves the dual purpose of audience expansion and content maximization.

4. AI and Advanced Tools: Artificial intelligence (AI) will play a pivotal role in content creation and editing, offering sophisticated tools for podcasters to streamline their production process. AI could assist in tasks ranging from transcription to content recommendation, making podcasting more accessible to a wider range of creators.

Brenden Mulligan is the founder of Podpage.com, a podcast-focused website platform that helps thousands of podcasters build and host incredible podcast websites.

30. TONI HENDERSON-MAYERS

"Livestreamers and social media influencers will discover the value of extending their brand to streaming channels."

TONI HENDERSON-MAYERS
North Star Stage & Screen

Livestreamers and social media influencers will discover the value of extending their brand to streaming channels.

In the same way TV died a slow death to cable, the death of cable became the same fate with streaming channels.

Businesses, livestreamers, influencers and all who are working to extend their brand should consider the move to streaming their content.

Toni Henderson-Mayers is an award-winning actress, director and livestreamer who took her worldwide book, "Wise Courtship," and branded herself through livestreams, podcasts, radio, TV and streaming. In addition to performing for TV, film and stage, Toni assists businesses in branding themselves or helping individuals become successful performers.

31. BRAD FRIEDMAN

"The podcasting industry and AI are projected to converge more significantly than ever, transforming the landscape of digital audio content as we know it."

BRAD FRIEDMAN
The Friedman Group, LLC

LIVESTREAM UNIVERSE

In 2024, the podcasting industry and artificial intelligence (AI) are projected to converge more significantly than ever. This integration may manifest in various innovative and groundbreaking ways, transforming the landscape of digital audio content as we know it. Here are some predictions:

1. **Personalized Podcast Recommendations**: AI will play an increasingly significant role in curating personalized podcast recommendations. Data about listener's behaviors, interests and preferences will be leveraged to generate tailored suggestions, increasing listeners' engagement with the platform.

2. **AI-Driven Content Creation**: With GPT-4 and subsequent iterations of AI, we can expect some podcasts to be entirely written and produced by amplifying human-like generated voices with varied tones and emotions. This will allow for streamlined content production, with less time spent on scripting and editing.

3. **Speech Recognition and Transcription**: The advancements in AI will improve speech recognition algorithms, making transcripts even more accurate and affordable. It opens podcast content to a larger audience, including those with hearing impairments and non-English speakers, enhancing accessibility and inclusivity.

4. **Interactive Podcasts**: With AI, podcasts will become more interactive. Imagine a scenario where you can ask questions to the podcast itself and get instant, accurate responses. AI-assisted podcasts will enable two-way interactions, breaking the fourth wall in podcasting, much like interactive TV shows today.

5. **Advertisement Matching**: AI will enhance targeted advertising within podcasts. It will analyze listener habits and demographics to determine the most suitable advertisements for specific time slots, leading to a more personalized ad experience for listeners and higher ROI for advertisers.

6. **Voice Cloning**: While controversial, the technology for creating ultra-realistic copies of human voices may become more prevalent. It would allow podcasters to produce content using the voices of famous personalities (with their consent), attracting a bigger audience.

While these predictions explore the positive power of AI, they also raise critical cautionary issues. Podcast creators and consumers must be aware of ethical considerations like data privacy, consent and the risk of deepfakes. Nonetheless, the convergence of podcasting and AI opens an exciting new chapter in the realm of digital audio content.

Brad Friedman is a digital marketing strategist and coach. He works with professional services providers and business owners to level-up their marketing to generate more revenue online. He is also an Amazon best-selling author of two books and the host of The Digital Slice Podcast, where he has been blessed to interview some of the brightest minds in the industry.

32. S. Chris Edmonds

"Without conscious attention, our workplaces can become as toxic as our social media."

S. CHRIS EDMONDS
The Purposeful Culture Group

Our society continues to suffer from incivility and divisiveness. With the election cycle ramping up here in the US, it's not likely that disrespect will go away. Without conscious attention, our workplaces can become as toxic as our social media platforms.

To create sanity and even — heaven forbid — respect and validation across your organization, business leaders must use every digital tool at their disposal to communicate openly, connect authentically, and clarify meaningful work to every team leader and team member frequently. Livestreaming is simple and impactful when it's a key piece of creating a work culture of respect AND results.

S. Chris Edmonds is the CEO of The Purposeful Culture Group. He's a speaker, author, executive consultant and musician. Chris and his team help leaders sustain uncompromising work cultures.

33. AMIR ZONOZI

"As social platforms move to more paid non-ad experiences, brands will invest more into marketing integration via product placement across sports, music and entertainment."

AMIR ZONOZI
Zoomph

We live in a world where attention is currency, content is commerce and audiences are assets.

I am confident in 2024 we will see viewership numbers continue to rise among sports properties as well as more brands entering sports partnerships and activating across broadcast, streaming and social to target niche audiences where they are.

As social platforms move to more paid non-ad experiences, we'll see brands invest more into marketing integration through product placement across sports, music and entertainment.

The value of these exposures, and understanding the audiences consuming the content, will be incredibly important; advancements in AI will help make tracking the ROI and impact easier for brands and properties.

Amir Zonozi is the CEO and cofounder of Zoomph, a partnerships media measurement and intelligence company across broadcast, streaming and social that supports customers like Golden State Warriors, Wasserman, NASCAR, Angel City FC, Invisalign and more. Amir also serves on the board of Women in SportsTech and Ally's Women's Sports Club.

34. KERSTIN OLETA

"More women will lead — using multiple forms of digital media to build their own brands — instead of attaching themselves under already established brands."

KERSTIN OLETA
Business Leadership Excellence Institute

My prediction for 2024 is that we will see a higher level of women leading — using multiple forms of digital media to build their own brands — instead of attaching themselves under already established brands. This will also bring stronger collaborations between high-level female leaders on a national and global level. It is the natural state and history of women to come together and support each other.

In 1954, Marilyn Monroe approached the owners of the famous LA nightclub Mocambo and promised to sit in the front row with other celebrities if owners allowed her friend Ella Fitzgerald to sing there, which they had originally refused to do. Marilyn's private influence moved Ella's singing from only being in small jazz clubs to the historic career that continues to live on.

Though women have always cheered each other on, I foresee this happening right in front of everyone's eyes. Women understand their worth of contribution using digital media in a different way and with a new confidence.

In the next year, this will create a consistent upswing for female leadership to scale to higher levels of business by linking arms courageously together centerstage instead of in the backdrop or next to their male counterparts.

I anticipate that we will see this prevalently in livestreaming, podcasting and industry events to drive more content creation and growth for online and in-person, female-owned businesses.

Kerstin Oleta is the CEO and founder of The Business Leadership Excellence Institute, a training system to learn communicating executive presence, leadership in business design and team management, and empowering female leaders. Kerstin works with business professionals and executives, performing keynote speeches and training for organizations such as Google, eBay and Amazon.

35. ROSS QUINTANA

"As brands embrace the shift from traditional channels like TV for Youtube to reach Youtube natives, ad spend from larger brands will drive an expansion of the creator community."

ROSS QUINTANA
Social Magnets

As brands embrace the shift from traditional channels like TV for YouTube (to reach YouTube natives), increasing ad spend from larger brands will drive an expansion of the creator community. We may see more mainstream programming move to YouTube as well as a decline in streaming services that have failed to deliver compelling content for years.

We could also see innovation in video commerce as video content moves to interactive video experiences bridging the worlds of Web3 and VR/AR to traditional video. Passively watching videos should progress to interacting with video to create alternate video paths as well as generate commerce through video and ecommerce innovations.

As video is the format of choice due to being the most lifelike in terms of density of communication and mimicking real-life experiences, look for a slow move in the direction of interactive video — something like a hybrid of website functionality in video and live video. This will create a natural on-ramp for more VR/AR content, so users don't have to choose one or the other.

As with all innovation there is a hype cycle and then the slow move to adoption and that gap can take years. People have been talking about video as a trend for more than five years and yet it is still growing and evolving.

The same can be said of VR and Web3, so don't look for overnight mass adoptions. For that reason, technologies that have been hyped but are progressing will see steady market adoption and refined iteration.

AI should accelerate all media categories as the startups that were in the hype cycle start to launch their products into the marketplace.

Ross Quintana is the founder of Social Magnets, a content design and social media management company. Ross has worked as head of social for Adobe Partners, consulted with Fortune 500 companies, startups and influencers around the globe, and is featured in multiple books on entrepreneurship, marketing and customer experience.

36. BRYAN KRAMER

"Email marketing newsletters are poised to dominate due to their unparalleled reach and effectiveness in engaging audiences."

BRYAN KRAMER
PureMatter

Email marketing newsletters are poised to dominate in the upcoming year due to their unparalleled reach and effectiveness in engaging audiences.

As privacy concerns grow and social media algorithms become more restrictive, email remains a direct and personal communication channel, allowing brands to connect with their subscribers more intimately.

The ability to segment and personalize content makes emails highly targeted, increasing their relevance and impact.

Bryan Kramer is the proud father of Human-to-Human: H2H. He's an executive coach and advisor, keynote speaker, two-time author, investor, virtual and in-person global keynote speaker, CEO of two companies and TEDTalker.

37. RENEE HASTINGS

"Because of 5G, our devices will be like superhero gadgets, helping us create and share even more amazing livestreams than we can imagine right now."

RENEE HASTINGS
Executive Help Now!

5G and Livestreaming: A Magical Gateway to Connection

Opening Act: The Buffer-Free Wonderland

Picture this: You're all set to watch an exciting livestream on your tablet or phone. But wait! Sometimes it takes ages to load, and the dreaded buffering kicks in. Fear not, for here comes our hero, 5G, the super-fast internet for your devices.

"So, this 5G thing," our tech wizard begins, "makes livestreaming way better. No more pauses or waiting. It's like experiencing your favorite live show, but now it's on your tablet or phone!"

Scene 1: Beyond Livestreams — A World of Livestream Magic

Hold on, it's not just about livestreams. 5G unveils a world of livestreaming magic for your devices. Virtual reality (VR) becomes your enchanted doorway to experience events in real-time. Imagine putting on special glasses and attending a live concert or exploring a distant location. It's a bit like magic!

"And for people who share live experiences online," our storyteller adds, "like that real estate agent we talked about, 5G helps them share things faster and in better quality. Imagine showing the world the beauty of where you are, live as it happens!"

Scene 2: 5G — The Superhero of Livestream Connectivity

But wait, there's more to this magical tale. 5G isn't just for fun; it's a superhero for important livestreaming moments too. Picture this: Livestreamers using 5G to broadcast from remote locations or share immersive experiences.

"And," our guide teases, "we might even see groundbreaking livestreams, like surgeries or exclusive events, happening with special glasses or cool technology."

"In simple words," our narrator sums up, "5G makes everything in our livestreams faster, clearer and more awesome. It's like upgrading to a super-speed internet that helps us share amazing live moments we couldn't do before. Exciting, right?"

Scene 3: The Future Beckons — A Glimpse of Tomorrow's Livestreams

Fast forward to a future where everything in your livestreams is even more awesome! In this super-fast internet era called 5G, livestreaming becomes a magical experience — no more waiting or pauses.

"In the future," our visionary storyteller shares, "people might use 5G to share live adventures. Livestreaming from incredible places, showing off amazing events they are a part of. Livestreaming becomes not just a broadcast but a real-time connection."

Closing Scene: Devices as Livestreaming Superheroes

"So, my prediction," our storyteller concludes, "is that in the future, because of 5G, our devices will be like superhero gadgets, helping us create and share even more amazing livestreams than we can imagine right now. It's going to be super exciting!"

Epilogue: Ray-Ban Meta Livestream Glasses — A Sneak Peek into Tomorrow's Livestreams

The chapter closes with a sneak peek into the future, introducing Ray-Ban Meta smart glasses. These glasses, equipped with a 12MP camera, built-in speakers and a five-microphone system, enable wearers to livestream, creating another exciting layer to the magic of 5G and livestreaming.

Our journey into the enchanting world of livestreaming with 5G is just beginning, and these smart glasses are another glimpse into the wonders that lie ahead.

Renee Hastings is the president and CEO of Executive Help Now, a virtual assistance service. She and her cohort of managers, specialists and executive assistants provide administrative support, video podcast production support and business consulting services to small business owners, content creators and busy executives.

38. Andrew Kavanagh

"Instead of thinking of AI as a threat, it'll be seen as an assistant making our lives easier, enabling us to look at tasks from a different perspective, expanding our possibilities and our vision."

ANDREW KAVANAGH
Digital Artist

I predict that AI will not only become more mind-boggling and advanced but will implement more features to make it easier — such as the ability to use speech to create AI art and AI-generated text summaries, articles and stories.

I also believe AI will be more focused on how it can make our lives easier with AI-powered virtual assistants. AI virtual assistants will come in the form of various sites and apps and will focus on helping us organize our business and daily duties, minimize tedious and repetitive tasks and suggest different perspectives and options for our business, marketing and creative projects, which will be very helpful for digital media and livestreaming project titles and descriptions.

Instead of thinking of AI as a threat, it will be seen more as an assistant that is making our lives easier and enabling us to look at tasks from a different perspective that will only expand our possibilities and our vision.

Andrew Kavanagh is a digital artist and Photoshop and Adobe Lightroom tutor. He does photo compositing and photo retouching in Los Angeles, CA. He's an Adobe Community Expert and an Adobe Express Ambassador. Andrew loves doing live video events and video tutorials that he shares on Facebook, LinkedIn and YouTube.

39. Professor Nez

"Content creators are going to rely less and less on analytics, strategies and gimmicks, and more on what electrifies their innards."

PROFESSOR NEZ
Professor Nez, LLC

A complete shift in the creator economy will commence. Content creators are going to rely less and less on analytics, strategies and gimmicks, and more on what electrifies their innards.

By creating from this epicenter, we will see a renaissance of content creation that is more personal, deeply vulnerable and ultimately more engaging.

Professor Nez creates educational content and consults brands and businesses on their content strategy.

40. NANCY DEBRA BARROWS

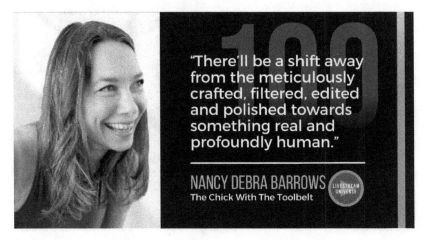

"There'll be a shift away from the meticulously crafted, filtered, edited and polished towards something real and profoundly human."

NANCY DEBRA BARROWS
The Chick With The Toolbelt

A powerful transformation has been taking place. A shift away from the meticulously crafted, filtered, edited and polished towards something real, "as-is," and profoundly human.

This organic authenticity, sharing the experience of LIVING life and being human is where the heart of genuine connection lies.

Today's social media users are more discerning than ever. They know disingenuousness when they see it. They're passing over the polished façade that has historically been presented in social media — a world where image often overshadows substance.

It's no longer enough to merely present a glossy exterior. Audiences crave real, unscripted moments. They want to see the people behind the brands and want them to be just as beautifully flawed as they are.

In the coming year, anticipate the continued ascent of platforms like TikTok and Instagram. In 2024, data indicates that Instagram is predicted to grow its number of users by 50 million to 1.4 billion. TikTok is poised to increase its user base by 8% to 900 million. These platforms thrive on creativity, quirky realness, entertainment and the allure of short, engaging content.

What truly fuels the fire of these platforms, however, are the influencers and creators who offer a refreshing departure from the manufactured image that has become all too common. Curated content is no longer king. The authenticity of these creators strikes a chord. It's not about being perfect; it's about being real.

Brands that embrace and commit to showing up honestly will be rewarded for bringing a unique, human aspect to their content that reaches distinct audiences and sparks engagement. Storytelling, behind-the-scenes and blooper reels take center stage. Authenticity in your brand's DNA — the basic building blocks that represent the core values and personality of the brand — is crucial.

In a world where the constant barrage of sales pitches has long lost its appeal, the key lies in content that evokes emotion — content that makes people laugh, gasp at the unexpected or even shed a tear. In these moments you become relatable and your brand reveals a heart that your audience can genuinely connect with. They see themselves in you and your brand.

As AI evolves to offer innovative, time-saving opportunities for content creation, it's the human touch that remains irreplaceable. It's the stories, the anecdotes and the candid moments that people seek out and come back to regularly. Content that resonates with communities, speaks to their hearts, their experience and shares their values. Content that makes them feel seen, heard and understood is what they respond to and reward with loyalty.

This shift toward authenticity isn't just a trend but a response to the modern consumer's desire — and demand — for transparency. People want to get to know YOU. A survey conducted by Stackla found that 88% of consumers prioritize authenticity when choosing which brands to support.

In 2024, this means bringing the consumer into the day-to-day, sharing behind-the-scenes footage, exclusive looks, candid moments and acknowledging blemishes and mistakes. Imperfection and error are what makes us human and relatable. Failure is no longer a four-letter word. It's about shining the light on the people that make up the company/brand. Shifting from a nameless, faceless machine to a family and community that people feel they know and are a part of.

In 2024, as your brand embraces real communication and authenticity, it opens doors to deeper connections and lasting loyalty. It's the willingness to share not just the gloss and glamor, but the real, unscripted moments that define your brand's journey.

It's the recognition that the goal posts have moved, and perfection is not the target; it's the genuine, human, relatable and imperfect nature of your brand that resonates most deeply because it's real and reflects the human experience.

In this age where organic, authentic content prevails, stand out, connect on a profound level and show the world who you truly are by #RadiatingReal.

Nancy Debra Barrows **transforms hearts and lives with her profoundly moving and inspiring keynote speaking and coaching. She empowers people to get unstuck and banish imposter syndrome with her Chick With the Toolbelt programs. As the Queen of Engagement, she's a five-time LinkedIn TopVoice, Top 50 Most Impactful People, Top 50 Most Inspired Connections, Top 250 Rising Stars and Influencers to Watch and three-time best-selling author.**

41. Dr. Aikyna Finch

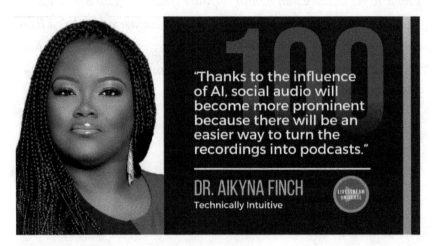

"Thanks to the influence of AI, social audio will become more prominent because there will be an easier way to turn the recordings into podcasts."

DR. AIKYNA FINCH
Technically Intuitive

My predictions for 2024 are as follows:

AI is going to make work and content creation much more effortless. It will become a way of life, like streaming or getting on the internet.

Because people will use AI to repurpose content, video will skyrocket again. Short- and long-form video content will grow equally in this season. People will be free to explore video now because of the influence of AI.

Social audio will become more prominent because there will be an easier way to turn the recordings into podcasts. People will also use AI to create the scripts, promotions and press releases for these podcasts using AI.

More online businesses will emerge due to the creation of membership and e-commerce sites in new industries. People will create websites, create product ideas and market these businesses using AI.

Dr. Aikyna Finch is a podcaster, coach, educator and TEDx speaker. She coaches executive, life and technology at the individual and group levels. She is the host of the Dr. Finch Experience® Podcast. In 2020, she started her tech coaching brand called Technically Intuitive®. In November 2022, she became a TEDX DeerPark speaker with her presentation on "The Benefits of AI on Social Media."

42. CHAD ILLA-PETERSEN

"The AI helpers will be like high-tech metal detectors, able to dig up forgotten tales buried in data. They'll sift through millions of reviews, posts and comments to uncover hidden story gems."

CHAD ILLA-PETERSEN
The Story Catcher, LLC

Here is a version that incorporates AI's role in a PG way: Get ready for an epic quest to rescue lost stories and craft new tales!

Picture this – it's the year 2024 and a massive storm just rolled through, tossing stories far and wide. Brands realize they must become storycatchers and set off on a mission to gather them back up.

This reminds me of one super windy day at summer camp. After lunch, we raced down to the lake only to find all the canoes had blown way out onto the water! My friend Tyler and I volunteered to swim out and tow them back. Let me tell you, that lake was freezing! The water was so cold it hurt. We wanted to give up right away and swim back. But we knew we had to rescue those canoes together.

Through chattering teeth, we pushed forward, grabbing ropes in our numb hands and pulling with all our might. After what felt like forever, we finally collected the last canoe and collapsed on shore. We were freezing but also feeling pretty heroic!

It's going to be the same for brands in the future. At first, catching all those stories blowing around will feel impossible and uncomfortable. But working together with their communities and trusty AI assistants, brands will slowly reel them back in.

The AI helpers will be like high-tech metal detectors, able to dig up forgotten tales buried in data. They'll sift through millions of reviews, posts, and comments to uncover hidden story gems. It will take teamwork, bravery and creativity.

Brands and their communities will craft new tales to unite people around purpose. Bit by bit, they'll build up a fleet of awesome stories!

So, get ready for an epic storytelling adventure! The future is all about grabbing oars and padding out across the winds of time to rescue forgotten stories before they're gone for good. It's gonna take grit — but the treasures we'll find will be epic!

Following a hardcore snipe hunt, botched pizzeria robbery, two years in a bear suit, and several moments atop a six-foot unicycle, Chad Illa-Petersen (The Story Catcher) discovered the compelling power of stories. Today, Chad helps you find and craft the stories you didn't even know you had to grow your brand, influence, and bottom line.

43. KEVIN KOLBE

"2024 could be a return to RI — Real Intelligence. You have a superpower that no AI will ever have and that's simply being you."

KEVIN KOLBE
Kevin Kolbe

I believe the continued rise of AI will see an opportunity for creators to be less dependent on AI and lean more into just being their own real selves. Yes, AI can be a great tool for ideas and curation, but the irony is AI still needs a human to tell it what to do.

You have a super-power that no AI will ever have and that's simply being you. So, 2024 could be a return to "RI" — "Real Intelligence."

Kevin Kolbe is a solo content creator, best-selling author and podcaster. He helps businesses, non-profits and other creators make an impact with online video.

44. JENNIE MUSTAFA-JULOCK

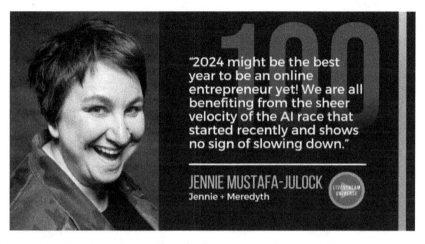

"2024 might be the best year to be an online entrepreneur yet! We are all benefiting from the sheer velocity of the AI race that started recently and shows no sign of slowing down."

JENNIE MUSTAFA-JULOCK
Jennie + Meredyth

I predict 2024 might just be the best year to be an online entrepreneur yet!

While we are still figuring out how to navigate the pandemic hangover, we are also coming out stronger than ever before. We are all benefiting from the sheer velocity of the AI race that started recently and shows no sign of slowing down.

And we're getting more powerful ways of boosting our signal and reaching larger audiences without breaking the bank (or our backs).

I'm personally most excited to watch what happens with social audio and course creation next. I'm onboard for a low-ticket revolution to shake up the coaching/consulting industry in a profound way.

Jennie Mustafa-Julock is the president of Jennie + Meredyth (formerly 'Coach Jennie'). Together, they are on a mission to change how you approach productivity, so you can actually accomplish more of what matters most to you... THIS WEEK...and every week after that. They run their growing two-woman empire from the road full-time in a beautiful 28' Airstream trailer with their trusty pup, Sadie.

45. Christoph Trappe

"If we embrace emerging technologies with wisdom, foresight and care, AI can take content creation to thrilling new heights."

CHRISTOPH TRAPPE
The Business Storytelling Show

The rise of artificial intelligence is opening new creative possibilities for content creators and marketers. As amazing as these AI tools are, it's vital we use them ethically and responsibly. Here are some predictions on how content creators can harness AI responsibly in the years ahead:

Content Outlining

AI tools are emerging that can analyze transcripts or written content and suggest helpful outlines. This could greatly assist creators in structuring podcast episodes, blog posts, videos and more. However, the outlines generated by AI should only serve as an initial framework. Creators should carefully review the outline, make changes to fit their unique vision and develop the final structure themselves. Leaning too heavily on AI-generated outlines runs the risk of cookie-cutter content that lacks originality.

Question Development

Many content creators interview guests for podcasts, videos and articles. AI can help generate interesting interview questions tailored to each guest and their expertise. But creators should take care to curate the computer-generated questions. Only use questions that align with the creator's values and avoid questions that could lead guests toward controversy simply to generate clicks. The responsibility is on the interviewer to sustain an ethical line of questioning.

Editing

AI tools are being developed that can analyze written content and suggest edits to improve clarity, flow and engagement. This can help creators refine their work. However, whenever AI proposes edits, creators should carefully review each change before approving it. Creators, not algorithms, should have the final say in the words, ideas and framing of their content. Using AI editing assistants responsibly means maintaining creative control.

Voice Cloning

Vocal synthesis technology enables creators to clone their voice with just a few minutes of sample audio. This opens exciting possibilities, like making podcasts even quicker to produce — as long as creators have written source content. But listeners could be misled if AI-cloned voices are not properly identified.

Expanding on Source Content

Creators can feed source material into AI systems to generate draft passages that summarize information and viewpoints. This can help accelerate the creation process. However, the creator should carefully fact check all computer-generated passages to ensure accuracy. If the AI includes biases or factual errors, the creator must revise the content. Creators must take responsibility for the truth and fairness of the final work — AI should only serve as an idea generator.

The common thread running through all of my predictions is that AI is a tool to empower human creativity, not replace it. Content creators should thoughtfully evaluate computer-generated suggestions but never forfeit their editorial judgment or creative vision. Maintaining human oversight and nuanced decision making while benefiting from AI's contributions — that is the formula for using these technologies responsibly.

Of course, AI capabilities are rapidly evolving. As these tools grow more advanced in the coming years, we must remain vigilant. Content creators, tech firms and regulators all have a role to play in ensuring AI is used ethically. If we embrace emerging technologies with wisdom, foresight and care, AI can take content creation to thrilling new heights. My prediction is that a large number of creators will do just that!

Christoph Trappe is the host of "The Business Storytelling Show, " blogs at ChristophTrappe.com and serves as director of content strategy at Growgetter.io. His latest book is "Is Marketing a Good Career?"

46. ALESSANDRA COLACI

"AI makes the ideation and production phases faster, while still giving room for people to make important adjustments."

ALESSANDRA COLACI
Humans + AI

AI will become an essential part of creator workflows. Content creation will become faster with AI for everything from script writing to creating multiple short clips for social media. Creators will be able to bounce ideas around with tools that will specialize in analyzing what might be the best topic to cover.

The most important thing is to still have a human element in the mix. People will always be the best at identifying which tweaks and messaging optimizations are needed for their specific audience.

AI makes the ideation and production phases faster, while still giving room for people to make important adjustments.

Alessandra Colaci is the founder of Humans + AI, a consultancy and podcast that helps businesses understand how to use AI for team efficiency and creativity. She has worked in digital marketing for over 15 years, primarily in SaaS and e-commerce.

Alessandra has managed marketing teams big and small and loves to share her passion for emerging tools that make teams up-level their output.

47. PHIL GERBYSHAK

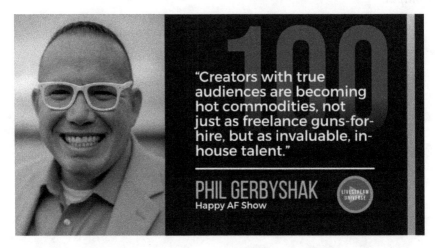

"Creators with true audiences are becoming hot commodities, not just as freelance guns-for-hire, but as invaluable, in-house talent."

PHIL GERBYSHAK
Happy AF Show

Listen up, folks: In an age where anyone with a keyboard can crown themselves an "expert" thanks to platforms like *Claude*, ChatGPT and *Charlie*, what's going to set you apart is your authentic personality and the tribe you build around it.

Brands aren't just looking for voices; they're looking for voices that resonate — voices that amplify their core values and can dominate conversations through both audio and video mediums.

Forget vanity metrics. Companies are wising up. They're scouting for creators who've built bona fide followings — not those who've simply bought followers to look the part. Why? Because the content battleground is more cutthroat than ever, and to stand out, you need more than just volume; you need caliber.

Creators with true audiences are becoming hot commodities within organizations, not just as freelance guns for hire, but as invaluable, in-house talent. They're the ones who can produce content that's not just engaging, but genuinely game changing.

Being a lone wolf won't cut it anymore. It's time to be part of a pack — a multitude of packs, to be exact. In this noisy world, your message must be laser-focused to hit its mark. Joining or contributing to relevant affinity groups or communities isn't just a nice-to-have; it's a must. These communities act as echo chambers that amplify your message, ensuring it cuts through the cacophony and lands squarely in the ears that need to hear it most.

Don't even get me started on quality! With the plummeting costs of top-notch hardware and software, there's zero excuse for mediocre audio and video. The bar's been raised, and it's going nowhere but up. If you want to play in the big leagues, you've got to look and sound the part.

And let's not ignore the elephant in the room — the great consolidation wave that's sweeping through the software landscape. Big SaaS platforms are on a shopping spree, gobbling up niche players not just for their tech, but for their unique voices and the loyal communities they've curated. It's like a game of high-stakes Monopoly, where the key to winning isn't just about owning the board, but owning the conversation.

So, what's the game plan? Forge your unique voice, build a die-hard community, produce killer content, align with the right tribes and never compromise on quality. Do this, and not only will you survive in this hyper-competitive arena — you'll thrive!

And creators: Make sure you have an integrated operational system in place to leverage all of this and show you off as the professional you are. From CRM to social media, to email marketing, to invoicing, you need to look like the pro you are if you're going to get the top dollar you deserve.

Meet Phil Gerbyshak, VP of growth at SpeakerFlow. Boasting 20+ years in sales and marketing, Phil is equally at home on Wall Street and Main Street. More than a sales expert, he's a relationship-builder and tech enthusiast. Author of "Zero Dollar Consultancy," he redefines sales norms. Off-duty, he captivates audiences with riveting keynotes. Looking to up your business game? Phil's the man!

48. NIEL GUILARTE

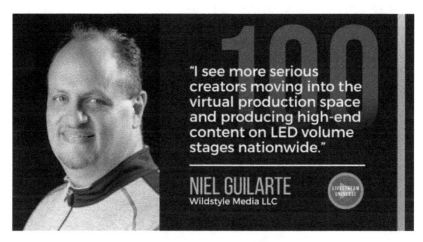

"I see more serious creators moving into the virtual production space and producing high-end content on LED volume stages nationwide."

NIEL GUILARTE
Wildstyle Media LLC

My prediction for content creators in 2024 is the rapid adaptation of AI tools specifically in the text-to-video editing space as well as the audio post-production space.

Additionally, I see more serious creators moving into the virtual production space and producing high-end content on *LED volume stages* nationwide.

Soon that technology will be available in more realistic ways and could possibly lessen the use of green screen technology as we know it. I currently direct virtual productions and this technology is the future!

Niel Guilarte is the founder of Wildstyle Media LLC, an award-winning media firm in Tampa, Florida. Niel is a podcast and business coach, filmmaker, podcaster and speaker.

49. JOIE GHARRITY

"As major media outlets move away from traditional programming, big brands will need to find other avenues to distribute their ad dollars."

JOIE GHARRITY
Joie G 113

As major media outlets move away from traditional programming, big brands will need to find other avenues to distribute their ad dollars.

Livestream content will become part of their formula for reaching their niche audience.

Joie Gharrity is the founder of Joie G 113 and the SWE Media Network. She empowers women entrepreneurs to be their own superstar by offering media opportunities with the Superstar Women Entrepreneurs Media Network, the livestream destination by, and for, women entrepreneurs. Each content creator provides superstar solutions in business, health, marketing, wellness, publishing, wealth and influence.

50. JOHN GIOVANNI PRETTO

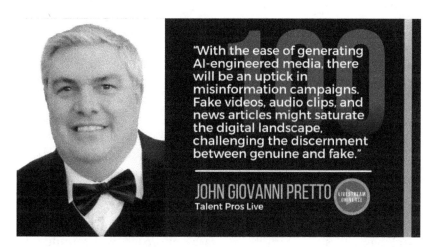

"With the ease of generating AI-engineered media, there will be an uptick in misinformation campaigns. Fake videos, audio clips, and news articles might saturate the digital landscape, challenging the discernment between genuine and fake."

JOHN GIOVANNI PRETTO
Talent Pros Live

The 2024 Presidential Election Year will undoubtedly witness a significant intersection between politics and AI. As technology becomes increasingly sophisticated, campaigns will harness AI tools for their promotional efforts. This could range from personalized advertisements and data analysis for targeted outreach to predicting voter behavior.

With the ease of generating AI-engineered media, there will be an uptick in misinformation campaigns. Fake videos, audio clips and news articles might saturate the digital landscape, challenging the discernment between genuine and fake.

Evolution of Media Creation Tools

Companies like *Descript*, which have been pioneers in AI-driven media creation, will face competition from industry giants like Adobe and Blackmagic. These companies are now integrating AI functionalities, offering a broader suite of tools for content creators.

The rise of text-to-video, text-to-3D and text-to-audio platforms indicates a trend where content creation will become faster, more efficient and more versatile.

AI Legislation

Given the potential risks and ethical dilemmas posed by AI, especially in media manipulation, it's likely that we'll see the introduction of AI-specific legislation. Both in the US and internationally, governments will begin regulating the development and use of AI technologies to protect citizens and maintain societal harmony.

Dominance of Blackmagic Designs

Blackmagic Designs, with its innovative approach, will continue its rise in the media hardware and software industry. By offering cost-effective and high-quality solutions, they will further solidify their reputation as invaluable partners for content creators.

Podcasting Continues to Crush

Podcasting's rise as a dominant medium can be attributed to several factors, many of which suggest that it has overtaken traditional terrestrial broadcasting in various aspects.

Let's explore these factors to understand why podcasting is primed to play an even more significant role in the future media landscape:

- On-Demand Listening
- Diversity of Content
- Low Barriers to Entry
- Integration with Smart Devices

- Monetization and Business Models
- Shift in Consumer Behavior

This list highlights just a few reasons why podcasting has and will continue to thrive as a primary source of media consumption.

John Pretto, often referred to as the "Grandfather" of streaming, has been a trailblazer in the digital broadcasting world. Starting a company in 1996 that harnessed proprietary streaming technology, John laid the groundwork for the streaming revolution we see today. Over three decades, John has been at the forefront of numerous technological evolutions, solidifying his legacy as a key influencer in the digital media space.

51. LOTTIE HEARN

"May 2024 bring some peace from online parsing, devices put down and return to a reach-out for real faces, family and friends."

LOTTIE HEARN
#LIVEWithLottieHearn

Having been forced to stop live and online connections due to 2020 MS body, brain and eyes saying "STOP!" — and the last three years wondering why, what and who I am, if I'm not #LIVEWithLottie virtually connecting and growing with you, my community?

Now moving to country life with little internet but nature abounding, it's been a tough slog away from Ireland online and back to IRL (in real life).

So, my 2024 prediction is less of that and more of a "please be kind to yourself" plea! May 2024 allow those who choose to, or have to, step away from our amazingly adaptable, business booming, all-encompassing ecosphere online. Remember to reminisce how wonderful the outside world is. Simply stunning, awesomely active and possibly more resplendent than when we forgot the world is actually beyond what's in our hands!

If able, may the 2024 world be helped by an ease off on-screen influencers, some peace from online parsing, devices put down and return to a reach-out for real faces, family and friends. Ideally in open-air healing nature spaces — before our data-farms destroy them all. Take time and be kind to yourself, before your body might shut you down too.

We can still listen live and learn from our on-screen leading lights when we can, but I now implore in 2024 to reconnect in real, not reel, life!

#LIVEWithLottie Hearn **was a regular show host and mentor since Blab days! #LadiesGoLIVE co-creator, #ConfidenceOnCamera author, collaborator, host, Predictions contributor, until MS health issues forced her to STOP in 2021. After 30+ years performing, presenting, producing, creating, coaching and mentoring #StageToScreen — she misses you, but is loving the real world again too!**

52. Desiree Duffy

"Writers are so often at the mercy of others and bogged down by processes that suck their time and energy. AI will give those writers time to focus on the creative side of writing."

DESIREE DUFFY
Black Chateau Enterprises

Writers will become friends with AI. Well, okay, maybe it will be more like frenemies, but as authors realize how powerful artificial intelligence is, and as AI improves and expands, savvy writers will benefit.

I foresee a future where authors will be able to create illustrations of their characters with AI and get the characters to look exactly like they "see them in their head." Writers will be able to produce work faster with AI assistants and GPTs designed to help with various types of editing, research, plotting and other tasks that often slow down the writing process.

I am optimistic about the future for writers because so often they are at the mercy of others, bogged down by processes that suck their time and energy, and often are expected to produce writing fast to meet consumer or a publisher's demands. AI will give those writers time to focus on the creative side of writing.

Desireé Duffy is the founder of Black Château, a marketing and public relations agency; Books That Make You, a Webby Award-winning multi-media brand that promotes books and authors through its website, podcast and radio show; and The BookFest® Adventure, a biannual online event that brings together booklovers from around the world.

53. MEIKO S. PATTON

"2024 will be the year of newsletters and the number one platform will be Beehiiv."

MEIKO S. PATTON
The Peripheral

2024 will be the Year of Newsletters. Newsletters saw a resurgence in 2023, and they will only continue to grow.

The number one platform will be Beehiiv. Beehiiv is built for growth, and it is from the same team that built Morning Brew.

In addition, Beehiiv has AI features built into the platform. You can use AI to write newsletter content as well as make images. We all know that A+ B = C. So, AI + Beehiiv = Clout in 2024.

Meiko S. Patton is the host of Ms. Beehiiv Podcast and owner of Beehiiv Newsletter Job Board. She helps YouTubers own their audience by starting a newsletter.

54. CHRIS STONE

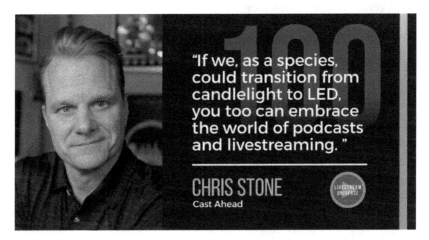

"If we, as a species, could transition from candlelight to LED, you too can embrace the world of podcasts and livestreaming."

CHRIS STONE
Cast Ahead

Navigating the Podcast and Livestream Galaxy: It's Not So Far Away as You Might Think...

As the two Tatooine suns set on traditional media, a new dawn beckons for those looking to venture into the realms of podcasts and livestreaming. For the Jedi entrepreneur master whose technological prowess might be described as, well, lacking, I hope this chapter can be your North Star — not your Death Star.

I. **The Modern-day Radio Renaissance: Podcasts**

In the corridors of digital sound waves, podcasts echo the sentiments of our age. They're the new-age radio, only without those pesky commercials about mattress sales.

Well, maybe they have some ads — but at least you can fast forward through them! Terrestrial radio is dwindling more and more — and the statistics are showing that even (gulp) children are listening to (and watching!) podcasts more than ever. The lines are also blurring between what a podcast is — audio or video.

Why venture into podcasting?

Symphony of authenticity: At the heart of every podcast lies an honest, unfiltered conversation — and one that you can control.

Your pace, your place: Be it during your audience's culinary experiments or their virtual cantinas, podcasts are their ever-ready companions. You reach them in their most intimate moments. Well, not the most intimate…

II. Livestreaming: The Art of Unedited Reality

Imagine capturing life with all its unpredictability, much like that time your banana bread decided to become banana soup. Livestreaming is the realm of raw moments and real emotions… and people are eating it up!

Quick guide for the analog soul: Platforms such as Amazon and YouTube Live are your trusted steeds. Mount them with confidence! The more you do it, the better you become at recording live and learning through the fumbles.

Eventually you'll have more recorded content to work with than ever before — that you can repurpose long after you've left the planet.

III. A Voyage to 2035: Bold Predictions

Podcasts in 3D: Imagine, your favorite podcast hosts, almost tangible, sharing wisdom right beside you! You can almost smell them! Ew. OK, we're not ready for 4D, yet.

Thought streams: A world where your thoughts echo in the livestream universe, sans words. Trippy, huh?

IV. For the Entrepreneur Whose Tech Tree Hasn't Bloomed Yet

Seek aid: The younger generation, often fluent in the language of tech, can be your guide. Offer help to them, and they'll learn to reciprocate.

Embrace knowledge: Course platforms and "YouTube University" are your sanctuaries. Every tech wizard once was a novice and had to start somewhere. Find someone that you resonate with and join their communities. You'll learn exponentially more and be able to create your own communities shortly thereafter.

Evolve: It's okay to falter. With every stumble, there's a lesson. With every lesson, there's growth. Use the force!

V. Pearls of Digital Humanoid Wisdom

Consistency is your compass: Regularity will lead you to your destination. Lead you, it will.

Forge connections: Engage with your virtual voyagers, cherish their feedback and steer your ship accordingly through this new galaxy.

The treasure of monetization: There are myriad ways to turn your digital journey into a trove of riches. Much like a stock portfolio, you'll need to have various streams of revenue to pull from as they will vary at times. If history has shown us anything, it's that evolution is inevitable.

If we, as a species, could transition from candlelight to LED, you too can embrace the world of podcasts and livestreaming.

So, as you turn this page, remember: The universe of digital wonders awaits, and you're destined to be its star. Don't be afraid to embrace things like video, virtual reality, ChatGPT, Generative AI, the Metaverse and Cliff Bars. You just might be energized to do better work.

May The Livestream Force Be With You!

Chris Stone boasts 25+ years with Sony Music, later founding Cast Ahead, a top-tier podcast and livestream consultancy. Chris has aided renowned podcasts and produced events for corporate giants. As a co-host of Amazon Live's "Dealcasters," he educates on content creation and Amazon's Influencer program. Notably, he's coordinated livestreams with Disney, IBM and MIT.

55. Bridgetti Lim Banda

"Livestreaming is not saturated; it continues an upward trajectory in the digital landscape."

BRIDGETTI LIM BANDA
B Live Media

An Era of Opportunities

Livestreaming is not saturated; it continues an upward trajectory in the digital landscape. Here's why it still remains an untapped goldmine:

Continued Growth: Livestreaming is on an unstoppable trajectory and an integral part of live events, Q&A sessions, educational content and e-commerce. Its value is undeniable, and its growth is set to accelerate as more businesses recognize the dynamic nature of live content.

Monetization Opportunities: Content creators can offer customized productions at a price point that big media simply can't compete with. Doors continue to open for creators to turn their passion into a sustainable income stream.

Emergence of New Platforms: New platforms and cutting-edge technologies are continuously entering the livestreaming scene creating opportunities for more innovative and immersive experiences.

Interactivity: Real-time audience engagement tools, such as live chat and live polls, continue to make livestreaming attractive. Two-way conversation enriches the viewer's experience, creates deeper connections and keeps audiences engaged.

International Growth: Livestreaming transcends borders; it's a passport to global audiences with the ability to impact a diverse and extensive community base.

Quality Improvement: Creators are continually raising the bar. It's no longer enough to go live; it's about delivering more engaging and compelling storytelling, improved audio quality and captivating content presentation.

Discoverability: As discoverability tools and algorithms evolve, content creators are finding innovative ways to attract their ideal audience.

The future looks promising for content creators who adapt to emerging trends and technology to get their slice of an ever-growing chunk of revenue.

Bridgetti Lim Banda, **an executive livestream producer, podcast host and author from Cape Town, South Africa, excels in creating engaging digital storytelling experiences.**

Her notable work on the Cape Town Water Crisis earned global acclaim, with features in print and broadcast media. Bridgetti is also a Global Goodwill Ambassador and an advocate for Invisible Disabilities.

56. Chris Krimitsos

"Video, video equipment and software fueled by AI will make high-quality video podcast creation more accessible."

CHRIS KRIMITSOS
Podfest

Video, video equipment and software fueled by AI will make the ability to create high-quality video podcasts more accessible.

Chris Krimitsos is the founder of Podfest Multimedia Expo, which is hosting its tenth annual conference in Orlando on January 25-28, 2024.

57. JIM FUHS

"Brands that dedicate their livestreams to delivering high-value solving experiences, not selling, will gain loyal followers."

JIM FUHS
Fuhsion Marketing

Livestreaming has become an incredibly powerful tool for businesses to connect with customers in real-time. However, the temptation is often to use livestreaming solely as a platform for sales pitches and product promotions.

While selling has its place, businesses that focus their livestreams on solving customer problems and delivering value will see greater success.

Here are some key reasons to use your livestreams for live solving rather than live selling:

Build Trust: When you focus on providing solutions instead of making sales pitches, you build credibility and trust with your audience. Viewers will come to see you as an expert guide who puts their interests first.

Boost Engagement: Content focused on solving problems and teaching skills attracts higher engagement and participation. Audiences want to learn and be helped, not just sold to.

Gain Insights: Listen carefully to the questions and issues raised during a live-solving session. You'll gain valuable insights into customer pain points to address.

Make a Difference: People want to do business with brands that make their lives easier. Live-solving shows you genuinely want to help versus just making a transaction.

Brands like eStreamly, Ecamm and others that dedicate their livestreams to delivering high-value solving experiences, not selling, will gain loyal followers who view them as trusted partners for the long term.

Jim Fuhs is the president of Fuhsion Marketing, a seasoned digital marketing consultant and a retired Marine LtCol with over 30 years of business and social media experience. Jim has a unique approach to marketing that fuses Marine Corps Leadership with next-level digital marketing, offering his clients a perspective that sets him apart from the rest.

58. ELIKQITIE

"More consumers will be moving over to paid search engines, such as Kagi, as the Google search platform is mostly a pay-for-play digital space."

ELIKQITIE
Write For You Ghostwriting

More consumers will be moving over to paid search engines, such as Kagi, as the Google search platform is mostly a pay-for-play digital space.

To get valuable content that is relevant to an internet search, internet users will need to have a paid search platform to receive search results that are filtered with personal preferences.

This way, search results will be relevant to the user's search and not based on how much someone paid for placement.

Lynn "Elikqitie" Smargis is an author, ghostwriter and writing coach who works with speakers, consultants and executives to create their first (or next) book to establish their credibility online. Lynn is an avid ideator who is currently producing three podcasts and writes between 2,000 and 8,000 words per day.

59. Marc Gawith

"Over the last few years we have seen many technology fads come and go. I predict that AI wins out and firmly becomes something that isn't a passing fad."

MARC GAWITH
Pictory

Over the last few years, we have seen many technology fads come and go. From livestreaming to AR/VR to AI. Each of these have had its day in the sun and we are firmly planted in the AI phase.

I predict that AI will be the one that wins out and firmly becomes something that isn't a passing fad. AI has become the way that companies continue to separate themselves from their competitors and it will continue to be a race to the top.

The developments and advancements that are happening in this space are very exciting. AI is already becoming a tool that early adopters need and see as a way to get ahead, but there is so much noise in this space from the companies that are, in effect, just wrappers for ChatGPT and other open-source *Language Learning Models* (LLMs).

Over the next six to 12 months AI will become super specialized. You will start to see companies specialize in solving a very unique problem. Think of it as going deep as opposed to going wide. It is an exciting time for AI and we are just scratching the surface.

Marc Gawith is the head of business development at Pictory.ai and the host of the Just Push Play podcast. He has 8+ years in the social video/live video and events space with a passion for video and AI. He has found his place leading BD, partnerships and sales at Pictory.

60. JOHN LARGENT

"In the senior care industry, the integration of digital media with patient care, disease management and advocacy will become standard practice."

JOHN LARGENT
Senior Care Movement™

In the senior care industry, the integration of digital media with patient care, disease management and advocacy will become standard practice.

With the senior population rapidly approaching more than 75 million by the end of the decade, livestreaming real-time product demonstrations, senior care consulting and education will become essential to serve the massive market of seniors and their families. Viewers will be able to purchase products or services directly through the livestream platform, reducing the steps from engagement to purchase.

In addition, medical device, pharmaceuticals and OTC companies will sponsor educational content that subtly integrates their products into the educational fabric of related topics. Branded podcasts that offer industry insights while highlighting a company's expertise will become more common.

YouTube will expand its role as an educational platform with companies offering detailed video tutorials, behind-the-scenes looks at product creation and educational content that ends with a call-to-action. YouTube's algorithms will likely evolve to better match educational content with user interest, improving opportunities for targeted marketing.

AI will become critical in personalized care plans and providing educational content to seniors and their families. AI will tailor learning modules, products and services while allowing virtual assistants to provide on-demand educational support and product recommendations.

John Largent is the host of Active Senior Lifestyles podcast and is the founder and CEO of the Senior Care Movement™, Senior Care Advisor University™ and owns Gameday Media Enterprises based in San Antonio, Texas. Senior Care Movement™ is a content-driven community driven by thought leaders, content creators and collaboration within the senior care industry.

61. BRIAN WALLACE

"We'll see a rise in more thoughtful, focused and targeted regional in-person events taking shape in the coming years. People crave human contact in person."

BRIAN WALLACE
NowSourcing

Since I've spoken at length about the future of LinkedIn being an important part of the future of digital in one's toolkit, I'll cover a different angle this time.

The event industry seems to be fraying on the edges. The global pandemic made it impossible for larger events to exist in person for years. That, and the fact that they were not equipped to exist as virtual events have many of these events hanging by a thread. Some have even ceased operations completely, while others have been acquired.

Many of us in the business world have made do without events for a few years. We are often tired of virtual events and are zoomed out.

My prediction: We will see a rise in more thoughtful, focused and targeted regional, in-person events taking shape in the coming years.

People crave human contact in person, especially given the loneliness many of us endured through Covid.

Brian Wallace is the founder of NowSourcing, an industry leading content marketing agency that makes the world's ideas simple, visual and influential. Brian has been named a Google Small Business Advisor for 2016-present, joined the SXSW Advisory Board in 2019-present and became an SMB Advisor for Lexmark in 2023. He is the lead organizer for TheInnovateSummit.com, scheduled for May 2024.

62. BEE SMITH

Visual podcasts will be the trend for 2024!

Listeners will increasingly desire to put a face to the voices they hear each episode.

Bee Smith is a bestselling author, speaker and educator pioneering mental health advocacy and inspiring change through the #BeeInspired digital brand.

63. ERIC HUNLEY

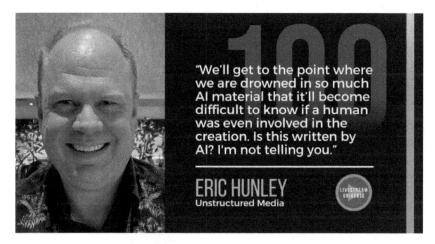

"We'll get to the point where we are drowned in so much AI material that it'll become difficult to know if a human was even involved in the creation. Is this written by AI? I'm not telling you."

ERIC HUNLEY
Unstructured Media

I think that 2024 will be a tumultuous year with a continuation of many unsettling challenges. It also will expand on many opportunities.

One big driver will be the continued rapid adoption of AI. We will get to the point where we are drowned in so much AI material that it will become difficult to know if a human was even involved in the creation. Is this prediction written by AI? I'm not telling you.

We also will continue to see an increased polarization between people both politically and culturally. This will become a strain on many creators as they see their audiences split into factions and face a continued onslaught of controversy, especially if they are political or have a strong viewpoint. This will get increasingly worse with it being a US Presidential Election year while people are still fighting over the previous results.

However, with all this increased tension, there is also a possibility for creators to grow and profit as many eyeballs will be glued to the screens and strife often results in more views.

As Robert F. Kennedy said in 1966, "There is a Chinese curse which says, 'May he live in interesting times.' Like it or not, we live in interesting times. They are times of danger and uncertainty; but they are also the most creative of any time in the history of mankind."

Eric Hunley is a serial YouTuber with four channels.

64. KAREN YANKOVICH

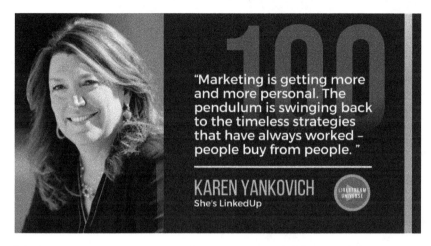

"Marketing is getting more and more personal. The pendulum is swinging back to the timeless strategies that have always worked — people buy from people."

KAREN YANKOVICH
She's LinkedUp

Marketing is getting more and more personal. The pendulum is swinging back to the timeless strategies that have always worked — people buy from people.

So yes, create awareness and visibility across platforms, but then come back on over to LinkedIn, where high-level relationships happen. Not 100 days of spamming your network, throwing spaghetti at the wall to see what sticks!

How about 10 per week? Outreach to 5-10 people each week, micro-targeted and researched, with the intention of having an actual conversation. That's where business is happening in 2023 and I see that only being stronger in 2024 and beyond.

We're tired of the noise and spam. Let's go back to LinkedIn for timeless, relationship marketing for our highest ticket opportunities. See you there!

Karen Yankovich is CEO of Uplevel Media, delivering profitable cutting-edge LinkedIn strategies, and host of the popular podcast "Good Girls Get Rich." Karen is an internationally recognized LinkedIn expert and consultant who is a genius at helping businesses use LinkedIn profitably.

65. PAUL RICHARDS

"The tools are finally available for fully remote productions to be set up and used for professional streaming."

PAUL RICHARDS
PTZOptics

Cloud based video production: The tools are finally available for fully remote productions to be setup and used for professional streaming.

This includes video switching, graphics, Comms, PTZ camera controls and more.

Paul Richards is the chief revenue officer for PTZOptics and creator of StreamGeeks. He is the author of several technology books, including "The Online Meeting Survival Guide," "Helping Your Church Live Stream," "Live Streaming is Smart Marketing" and "The Unofficial Guide to OBS." Richards teaches over 50,000 students on UDEMY on live video production, mobile streaming and more.

66. KATARINA ANDERSSON

"I foresee a continued emphasis on personalized, honest communication in marketing, community building and establishing direct connections to customers through personal platforms."

KATARINA ANDERSSON
Grapevine Adventures

In 2024, I foresee a continued emphasis on personalized and honest communication in marketing, community building and establishing direct connections to customers through personal platforms such as websites, newsletters and membership clubs.

Live video will continue to grow, in this context, as a way to capture the audience and interact with them directly. Interacting and creating a dialogue with followers and viewers will, in my opinion, be increasingly important.

In the wine industry, I believe it will be crucial for wine brands to better understand who their actual audience is and concentrate on that to create brand trust. Jumping from one trend to another will not help the wine sector.

Millennials and Generation X have the most buying power, despite Gen Z being the talk of the hour. It will be crucial to focus on building and nurturing your community and prioritize your own foundation, such as your website or newsletter.

Social media channels have proven to be increasingly unstable, which makes it even more important to invest in a platform that you own.

I believe it can be an added value for more and more wineries to create their own wine club program to sell directly to their customers (DTC).

Additionally, with stricter privacy regulations, transparency and the protection of your community should be a top priority.

Katarina Andersson is a wine writer, content strategist for wineries, translator, sommelier and livestreamer. She was born in Sweden but has lived in Florence for 20 years. She has a background in academia and a Ph.D. in history from EU's European University Institute in Florence. Katarina founded WinesOfItaly LiveStream in 2015 which was a weekly show up until 2019 and then became seasonal.

67. LAURA CLAPP DAVIDSON

"We'll see a movement towards smaller, more personalized creator conferences and events and a move away from large-scale shows."

LAURA CLAPP DAVIDSON
Shure

I predict we will see a movement towards smaller, more personalized creator conferences and events and a move away from large-scale shows.

People are craving connection, knowledge and skills, and this platform seems to hold up to those needs.

Laura Clapp Davidson heads up the market development team at Shure while being a mom, writing songs and hosting her weekly podcast, Song 43.

68. WIN CHARLES

Podcasting still is the best choice. YouTube shows are going up in popularity.

Win Charles **is a disabled rights educator.**

69. Dr. Stuart Buchan

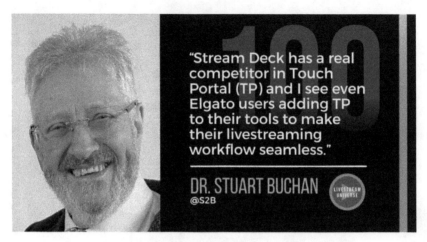

"Stream Deck has a real competitor in Touch Portal (TP) and I see even Elgato users adding TP to their tools to make their livestreaming workflow seamless."

DR. STUART BUCHAN
@S2B

There seems to be a move by some creators to transfer their livestreaming and on-demand content over from YouTube to Nebula. I can see that expanding, as it appears to be a better experience for both creator and consumers of their content.

Demand for 4K is on the increase, so default settings for video creation will be 1080p, rather than 720p. Facebook only allows members of their Level Up Program and managed partners to stream in 1080p, but this could well be opened up. The Metaverse will drive up demand. Creators who educate their audience will be the target for Facebook and YouTube in particular.

The issue with increasing video resolutions is storage and server processing. Server farms and data centers are gearing up for extra demand. Server-based livestreaming providers like Streamyard have creators who are keen to have multiple camera views, extra features that are akin to popular TV broadcasting standards or even OBS controls!

Even having 1080p instead of 720p as the base resolution for a paid subscription is a challenge. There will be more creators livestreaming from mobile apps, whether they are on iOS or Android devices. Most though will be (re)creating their own studios.

Stream Deck has a real competitor in Touch Portal (TP) and I see even Elgato users adding TP to their tools to make their livestreaming workflow seamless.

As an aside, I also envisage that more creators will add book authoring to their income streams!

Stuart Buchan has been livestreaming since 2006. He is a retired Chartered Engineer and now provides mentorship and tech support, plus creates and produces livestreams for his local church.

70. KATIE HORNOR

"More people will pay for online programs that are pre-recorded in smaller 5-10-minute chunks over live interactive trainings or long webinars."

KATIE HORNOR
KatieHornor.com

I predict that 2024 will find more people willing to pay for online programs that are pre-recorded in smaller 5-10-minute chunks (over live interactive trainings or long webinars) as our society once again prioritizes time around in person learning and activities.

Katie Hornor is one of only 0.002% of business owners in the world who are women, running a company that grosses over $100k/year while also being a full-time mom. Named as one of the Top 10 Entrepreneurs to Watch in 2022 by FOX, CBS and NBC, she's a ten-time best-selling author, business coach, an international speaker and the host of a top 2.5% global podcast. And, she loves flamingos.

71. TERRY BROCK

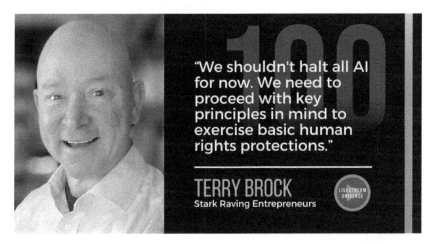

"We shouldn't halt all AI for now. We need to proceed with key principles in mind to exercise basic human rights protections."

TERRY BROCK
Stark Raving Entrepreneurs

AI is going to continue to grow (of course!!). However, there will be increased concern for privacy, copyright and more. We shouldn't halt all AI for now (as some have voiced).

We need to proceed forward with key principles in mind to exercise basic human rights protections and other measures. Content Creators have a huge possibility for growth, and I predict the opportunities will get bigger and better all the time.

Terry Brock is a marketing advisor and professional speaker (Hall of Fame member). He works with clients to help them deploy the right technologies to build solid, mutually-beneficial business relationships by leveraging AI and other technology.

72. KAREN GRAVES

"The Digital Marketing that will make the most impact will be that which is full of integrity and client-focused."

KAREN GRAVES
Karen Graves Coaching

My prediction: Digital marketing that will make the most impact in the upcoming year will be the marketing that is full of integrity and client-focused.

For a few reasons:

1) People are ready to return to community. They are craving human connection and interaction.

2) More live events are returning. Leveraging digital marketing to drive people to connectivity and collaboration will be the marketing that is most noticed.

3) Authenticity is always the charge. However, the rapid upsurge of technological advances (I'm looking at you AI) and new online entrepreneurs leave room for large volumes of bad actors. This is causing increased skepticism and distractions.

Fighting for the attention of consumers in a crowded and highly competitive marketplace when there is the potential for strategically placed misinformation by opportunists is turning the online space into the Wild Wild West. Any opportunity to show that your brand is genuine, ethical and customer/client-centric should be built into your marketing strategies.

Karen Graves uses 25+ years of experience in sales to help coaches, consultants and experts who sell offers $3k - $100k + get more yes with more ease with the Frictionless Yes Method.

73. Jeffrey Fitzgerald

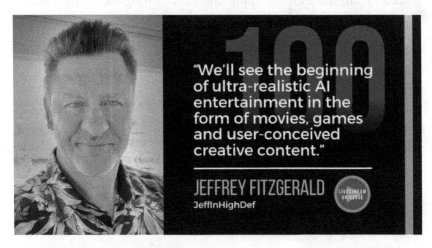

"We'll see the beginning of ultra-realistic AI entertainment in the form of movies, games and user-conceived creative content."

JEFFREY FITZGERALD
JeffInHighDef

What else could a prediction in 2024 be centered around, but AI?

It's the most culture shifting transformational technology since the dawn of the internet itself. With every new use case, AI stands to not just reshape every industry, but truly disrupt it to the core.

In the near future, no part of vocational life will be void of using thought computing as a framework, if not doing the work. Behind the scenes AI has already been at work for years in film, sales and even your iPhone photos. The ever-improving quality of ChatGPT's intelligence for writing along with *MidJourney*'s ability to craft near photographic images sets the stage for this year's prognostication.

We know AI is the change agent of our time and thought computing will one day be the only computing there is. But just how far can it go?

In the world of entertainment, I see something beyond amazing.

Imagine coming home from work, grabbing ice cream and getting comfortable on the couch. Tonight, it's Netflix and chill, so the search begins. Only you are in the mood for a movie you just can't seem to find. You want a comedy. But Kevin James and Mitch McConnell in 'A Capitol Mistake' isn't your thing. You want a romance. But Hallmark's 'Another Stranded Christmas' seems unappealing. So you just sit there fighting the remote for an endless supply of movies — only to pass on every one of them.

But what if it was Sayflix? The world's first generative AI for home entertainment. You simply begin to describe the kind of film you want to watch, add some story reference points and sit back.

Picking up the remote again, you shut your eyes and speak up. "Sayflix, I want to see a zany 90 minute comedy. About an international spy named Troubadour, who tries to protect the queen of England from the mob at a professional wrestling tournament. The main character drives an MG and is witty and sly. But he gets distracted by donuts. His pet Alfred is a super smart crime-solving parrot that likes to play piano."

You then say, "Make this movie in the style of Mission Impossible meets The Greatest Showman and surprise me with cameos by Winston Churchill, Sean Connery and Dua Lipa."

Click. And quite literally, the lights go down, the movie logo appears and the 'youvie' starts immediately. All of the scenes look as if Hollywood spent millions on production. The juxtaposed story lines — as weird and unconventional as they may be — all come together just like it was ready for the Oscars.

You could even say, "Put me in the movie as Trubador's best friend" and all of a sudden you are on the big screen. In love with the coolest spy in Europe solving mysteries together. Or what about those moments where you're watching a real movie classic that's just too long. "Sayflix, I've got to get to work. Show me Lord of the Rings in 30 minutes." You just saved yourself three hours, as AI decides on what scenes are meaningful and self creates the bridging scenes to retain consistency.

The speed of processing power is the only barrier to this happening, even today. But with the emergence of such profound advancements as *quantum computing*, predictions like Sayflix are entirely possible.

Quantum-powered AI could reduce the time of content creation and rendering to fractions of a second. Google says this about the kind of processing speed we are talking about: "Quantum computing is a new generation of technology that involves a type of computer 158 million times faster than the most sophisticated supercomputer we have in the world today. It is a device so powerful that it could do in four minutes what it would take a traditional supercomputer 10,000 years to accomplish."

No, quantum computers won't be on the shelves at BestBuy in 2024. But what could be the start? You will see the beginning of ultra-realistic AI entertainment in the form of movies, games and user-conceived creative content. Whether or not Kevin James and Mitch McConnell are in it, no one can really say...

JeffInHighDef *has been in media, TV, technology and content creation for over 40 years. He is a three-time 2020/2022 Emmy Award-winning video producer and the creator of The Digital Green Room, a Facebook community for media creatives.*

74. BETH GRANGER

"We will go through a period of disruption due to AI as we figure out if, and how, to use it appropriately and ethically."

BETH GRANGER
Intrepid Social

I think we will go through a period of disruption due to AI as we figure out if, and how, to use it appropriately and ethically.

After that, I am hopeful it will be used for what it is really good at, such as doing repetitive tasks or analyzing large amounts of data.

Beth Granger is a trainer, consultant and speaker who works with organizations and individuals, taking them from confused to confident on the platform. She is also an Exactly What to Say Certified Guide, helping people identify and elevate life's critical conversations. LinkedIn recognizes Beth's skills, frequently asking her to beta test new features.

75. WÁGNER DOS SANTOS

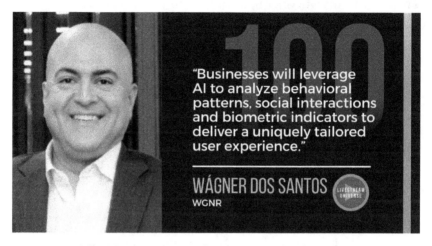

"Businesses will leverage AI to analyze behavioral patterns, social interactions and biometric indicators to deliver a uniquely tailored user experience."

WÁGNER DOS SANTOS
WGNR

The Convergence of Hyper-Personalization, Decentralized Platforms and Ethical Marketing

Predictions for 2024:

Hyper-Personalization Takes Center Stage

By 2024, the era of generic marketing will be long gone, replaced by AI-driven hyper-personalization. With 4.9 billion people using social media in 2023, expected to rise to 5.9 billion by 2027, the data available for personalization is immense. Businesses will leverage AI to analyze behavioral patterns, social interactions and even biometric indicators to deliver a uniquely tailored user experience. This will not only enhance customer engagement but also significantly improve ROI, which can already reach up to 700% with effective SEO strategies.

The Rise of Decentralized Platforms

As privacy concerns grow, decentralized platforms will gain traction. These platforms will offer users full control over their data, responding to increasing demands for online privacy. In 2023, 43% of consumers indicated they would switch brands following a bad privacy experience. Decentralized platforms will become the new norm for those who prioritize privacy, thereby disrupting the current centralized social media giants.

Ethical and Inclusive Marketing

The call for more ethical and inclusive marketing will be louder than ever. Brands will shift from merely minimizing their environmental impact to actively promoting important causes. In 2023, brands like Domino's and Diageo already started to critically assess their media placement choices to better connect with diverse audiences. By 2024, an authentic commitment to causes and inclusivity will not just be a 'nice-to-have' but a 'must-have'.

The Evolution of Email Marketing

Despite being one of the oldest digital tools, email marketing will experience a renaissance. As of 2023, more than 347 billion emails are sent and received every day. Businesses will integrate advanced AI algorithms into their email marketing strategies, pushing the ROI even higher than the current $36 for every dollar invested.

Gen Z Takes the Lead

Gen Z, the first generation to have fully grown up with the internet, will dictate the platforms and types of content that are relevant.

Brands will need to adapt to dynamic and highly visual platforms to meet their ever-changing needs.

By 2024, the digital landscape will be a complex tapestry of personalized experiences, decentralized platforms, ethical marketing practices and old-but-gold strategies like email marketing, all led by the whims of Gen Z.

Wágner dos Santos, **president of WGNR,** *blends brand psychology and storytelling in digital marketing. A pioneer since 15, he launched early online platforms and collaborated with giants like Google. Hosting "Wagner Live," he candidly discusses business. Awarded and recognized, he's been an industry leader, public speaker and socio-political activist.*

76. JESSICA KUPFERMAN

"Podcasters will learn to heavily utilize software like CastMagic to help with social media, transcription and show notes."

JESSICA KUPFERMAN
She Podcasts

My prediction is that many podcasters will attempt to incorporate YouTube into their publish channels and will give up within a few months, because it doesn't really make sense for us traditional podcasters yet.

I also think podcasters will learn to heavily utilize software like CastMagic to help with social media, transcription and show notes, because it does hours of work in 30 seconds.

There will be more AI competitors just focused on podcasting, too.

Jessica Kupferman is an innovative podcast expert and dynamic speaker, known for her wit and business acumen. As a seasoned comedian and writer, she engages audiences and empowers entrepreneurs with her rule-breaking strategies, making waves in the digital media landscape.

77. CHRIS CURRAN

"The era of online social authoritarianism is ending. It's time to start growing your audience and income using the next evolution of the social internet (Nostr) and money (Bitcoin)."

CHRIS CURRAN
Podcast Engineering School

Creators will receive more financial support than ever directly from their fans by utilizing decentralized protocols like *Nostr* and Bitcoin.

A major shift has started: Creators are shifting away from centralized platforms that can take away their audience for any made-up reason with zero accountability, toward new technologies and protocols where the creator's audience can never be taken away from them for any reason.

The era of online social authoritarianism is ending; it's time to start growing your audience and income using the next evolution of the social internet (Nostr) and money (Bitcoin).

Chris Curran is the founder and lead instructor of Podcast Engineering School where he teaches individuals how to earn a great living from home producing podcasts for clients.

78. Louise McDonnell

"The future of digital media will be shaped by those who can seamlessly integrate AI tools into their strategy without losing the human touch."

LOUISE MCDONNELL
SellOnSocial.Media

In the ever-evolving landscape of digital media, the 2020s witnessed a paradigm shift that few could have predicted. Among the plethora of innovations, AI tools, particularly ChatGPT, emerged as a game-changer. The rate of adoption of ChatGPT was nothing short of phenomenal.

To put it into perspective, while industry giants like Netflix took 3.5 years to amass a million users, Facebook took 10 months and Instagram took 2.5 months, ChatGPT shattered all records by reaching the same milestone in a mere five days. This unprecedented growth signifies not just the technological prowess of AI tools but also the changing dynamics of user preferences and needs.

The digital audience today is more informed, discerning and hungry for real-time, personalized content. AI tools like ChatGPT cater to this demand by offering instant, tailored responses, making them an invaluable asset for brands, influencers and content creators.

However, with great power comes great responsibility.

As we step into 2024, the challenge for brands and content creators will not just be about leveraging these AI tools but doing so in a way that retains authenticity. The digital space is already saturated, and with AI tools becoming mainstream, the competition will only intensify. The key will be to strike a balance — harnessing the efficiency and consistency offered by AI while also producing unique, authentic content that resonates with the audience.

In essence, the future of digital media will be shaped by those who can seamlessly integrate AI tools into their strategy without losing the human touch. The brands that can master this delicate balance will not only stand out from the crowd but also set new benchmarks in the realm of digital engagement.

An award-winning social media trainer and strategist, Louise McDonnell is the founder of SellOnSocial.Media and is a four-time best-selling social media author. She works with coaches, consultants, authors and online experts to leverage social media to drive business growth by using social media organic and paid strategies.

79. JONATHAN TRIPP

"Social media platforms will further leverage personalization and engagement using AI to better understand user preferences and needs."

JONATHAN TRIPP
Jonathan Tripp

Social media and messenger app usage will keep on growing in significance over the next year outpacing traditional media channels. Younger users will shift from using social media primarily for entertainment to using it more for connecting, learning and shopping.

Social media platforms will further leverage personalization and engagement by leveraging AI to gain a deeper understanding of user preferences and needs.

Jonathan Tripp is an experienced marketing professional with a strong focus on higher education and emerging target markets. He is a guest speaker and consultant with expertise in analytics-driven strategies, entrepreneurial ventures and event-based marketing.

80. JOHN KAPOS

"While the integration of Augmented Reality (AR) into marketing campaigns is on the horizon, a critical aspect to consider is the risk of AR experiences being perceived as overwhelming to consumers."

JOHN KAPOS
Perfection Chocolates

Augmented Reality (AR) technology represents an exciting trend that is poised to play a significant role in the business marketing landscape in 2024. This technology offers brands unique opportunities to connect with their customers and craft memorable brand interactions.

While the integration of AR into marketing campaigns is on the horizon for 2024, a critical aspect to consider is the risk of AR experiences being perceived as mere gimmicks or overwhelming to consumers. Brands must focus on ensuring that their AR offerings deliver genuine value and enhance the overall customer experience, rather than merely serving as eye-catching but superficial additions.

For example, an AR app will allow customers to see how their order is being made and packed, and how it would look presented as a gift, and even on the table ready to eat.

AR is indeed a trend worth monitoring closely in 2024, but its suitability may not be universal yet.

John Kapos, aka ChocolateJohnny, is a third generation Chocolatier and CEO (Chocolate Eating Officer) of his iconic family business Perfection Chocolates, Est 1939. John has taken his family business to a global brand using digital marketing and all aspects of social media. John is also an international keynote speaker. John helps, teaches and educates small business owners, corporate companies and entrepreneurs who need to expand, innovate and stand out of the crowd using social media, livestreaming, TikTok, brand and their voice.

81. TERRY JOHNSON

"Personal branding will give businesses a competitive edge online, helping them stand out in a crowded digital space and attract a loyal audience."

TERRY JOHNSON
Terry Johnson Online

As 2023 comes to a close, we see that the digital landscape is rapidly changing and evolving. This emphasizes the need for a personal brand that can stand out online.

We are discovering that personal branding is essential for content creators, entrepreneurs and small business owners, especially those just starting their online journey. This group will need personal branding to differentiate themselves online and attract a dedicated following.

Creating a personal brand can help solve the problem of being easily found online, especially with the increasing popularity of generative AI tools. While extremely helpful for creating content, these tools have flooded the digital space as of late. As a result, the competition for visibility has become more intense.

Personal branding will give businesses a competitive edge online, helping them stand out in a crowded digital space and attract a loyal audience. This single factor could help them achieve online business success in 2024 and beyond.

Terry Johnson is a marketing strategist who assists small business owners in optimizing marketing strategies for growth online. She focuses on different subjects like social media marketing, content marketing and branding to stay updated on the latest trends in the creator economy. Terry offers varied insights on crafting effective digital marketing strategies for business success online.

82. Melanie Falvey

"Content creators who opt to entrust every facet of video production to AI will undoubtedly boost their output but risk losing the invaluable connection with their audience."

MELANIE FALVEY
Expert Channel TV

As AI continues to gain widespread prominence, a noticeable divide is emerging among content creators. Those who opt to entrust every facet of video production to AI will undoubtedly boost their output but risk losing the invaluable connection with their audience.

We are already beginning to see some creators and marketers showing up as avatars (even without disclosing the fact that they are using AI tools), and this, in my opinion, is going to backfire.

On the other hand, livestreamers, podcasters and video content creators who persistently invest their time and hone their communication skills in their craft will reap the rewards of an appreciative and discerning audience. The personal connection forged through hands-on content creation, from the way you engage with your viewers through the lens to your real-time interaction during live podcasts or videos, remains irreplaceable by AI.

Personal brands, creators and channels who want to build trust with their audience will still need to show up in their videos or podcasts. I view AI as a potent research and brainstorming tool with its unique value, yet the art of human connection and communication requires our direct involvement in video creation. I firmly believe this is a positive development for our industry, empowering the most dedicated creators to stand out and thrive in a crowded landscape.

In a world of ever-increasing content saturation, the quest for quality becomes more and more important. Today's viewers and listeners are discerning, spoiled for choice and have begun to favor substance over sheer volume. This shift presents a golden opportunity for both aspiring hosts and seasoned professionals alike. It's a chance to invest in honing their skills, deepening their communication prowess and elevating their brand to new heights.

What's intriguing, however, is the distinct trend I am witnessing in VOD (video on demand), *SVOD*, *AVOD* and *CTV* platforms. Here, I see a striking disparity when it comes to content quality, as certain platforms prioritize quick advertising revenue over producing top-notch content.

While there's undoubtedly a demand for easily digestible, light-hearted content — a demand that's been around and will persist — creators and producers committed to delivering high-caliber material must conduct diligent market research before engaging with these platforms, ensuring the perfect alignment between their offerings and the platform's audience.

Melanie Falvey, once camera-shy, has become an influential media force, spearheading Expert Channel TV. Named Most Outstanding International TV Presenter, she is an award-winning visibility and brand authority strategist, video confidence coach, TV host, producer and mentor. She designs premium brands for entrepreneurs, coaches and brands through strategic visibility and authority building.

83. LONA DeRIEUX

"YouTube will become more popular for education and selling in both homeopathic and allopathic ways of thinking."

LONA DERIEUX
Happy Healing Inc

My Prediction for 2024 from my authentic self: YouTube is such a great platform for learning.

From an entrepreneurial point of view, I see YouTube becoming more popular for education and selling in both homeopathic and allopathic ways of thinking. I would call it the "Go to for business, emotional and physical wellness tips."

More people will be shining their light on all platforms since the uncomfortable people took action to become more comfortable. This action will help even more people to recognize it is time for all of us to shine. The spotlight will provide solutions for the best and highest good of us all.

Lona DeRieux is a doTERRA wellness advocate and energy worker. She is a certified Emotion Code (EC) and Body Code (BC) practitioner who specializes in a revolutionary energy balancing system.

Lona has a unique approach to the use of essential oils by identifying the emotional aspects of an illness, combined with muscle testing, to select the most appropriate oil(s) for each client.

84. JS GILBERT

"We will be turning more to actual known trusted circles and away from peer reviews, ratings and critiques, and back to professional critics, reviewers and established experts."

JS GILBERT
Business Problem Solving

I work heavily on credible advocacy marketing and I believe we will be turning more and more to actual known trusted circles and away from peer reviews, ratings and critiques, and back to "professional" critics and reviewers and established experts.

I suspect the continued explosion in influencer marketing will continue for another 18 months or so and then will die down pretty quickly. Influencers continually break the trust contract and this (along with deep fakes and AI entities) will become their major downfall.

Trusted, early adopters who are truly knowledgeable on the topics they speak about will become the new influencers. These individuals will tend to speak more about both the positive and negative aspects of something, as opposed to the current crop that seems to praise everything, as they direct you to their affiliate links.

I also think that advertising as a sustainable income stream for social media sites will also go bye-bye. More and more companies are realizing that online advertising is extremely corrupt and has been for a long time. Programmatic (real time online ad bidding) is full of bad actors, which is currently causing advertisers to retreat and bringing the overall cost of ads down by 50-80%.

Thus, social media sites can no longer exist on this meager income.

I expect that social media sites will soon start adopting systems akin to Patreon, in an effort to develop both a reliable, sustainable income stream and as a way of keeping the bad actors, bots, etc., out. Change will appear as a series of incremental changes, as everybody tries to adapt to doing business in the age of AI.

A final note is that two-way live online communications, such as that we see employed by Zoom, as well as other humanizing forms of online communications, will continue to grow, although there will be additional caveats in play. It will be incumbent upon us all to keep fully up to date.

JS Gilbert has worked as an advertising creative and broad-based business problem solver for decades. He is currently working on advocacy marketing, video, including live, and workplace compliance issues.

85. NANCY MYRLAND

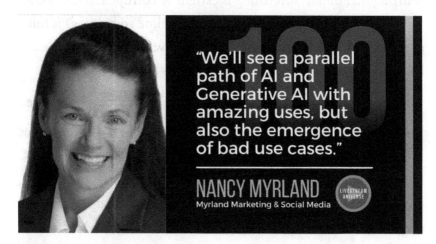

"We'll see a parallel path of AI and Generative AI with amazing uses, but also the emergence of bad use cases."

NANCY MYRLAND
Myrland Marketing & Social Media

LIVESTREAM UNIVERSE

I have a new addition for 2024. First, however, I remain bullish on voice and video, with the reminder that what I call "hand-held video" and "desktop video" on your nearest devices are very effective because they show your skills and your personality, which is appealing to other human beings who are looking for ways to connect and build community with you.

In 2024, I will add that we will be seeing more of AI in many ways. We will see a parallel path of AI and Generative AI with amazing uses, but also the emergence of bad use cases as well. We will see guardrails and legislation evolve that will attempt to protect the world, but there will always be bad actors that will take the "magic" and make it not so magic.

Whatever we can imagine as being seemingly impossible is now up for grabs as the opportunities and the technology are helping those of us who are interesting become much smarter and more efficient at what we do, which can ultimately benefit our clients and customers.

Nancy Myrland is a marketing and business development advisor, specializing in content, social and digital media. Her clients call her for strategy and planning, for all of their LinkedIn training needs, podcasting consulting and launches, and Zoom and virtual presentation coaching. She blogs at the Myrland Marketing Minute Blog and hosts 2 podcasts, Legal Marketing Minutes and Legal Marketing Moments.

86. MIKE GINGERICH

"Adoption of AI among the general public will not increase significantly from a pursuit standpoint, but they will be using it unknowingly in the software they use daily."

MIKE GINGERICH
Mike Gingerich Global

2023 has been the year of Artificial Intelligence and 2024 will be no different... except for the AI focus areas. What I mean by this is that the rush to develop AI has created the need for regulation and tools that protect from shady and fake AI.

So, while AI will continue to gain adoption in marketing and software like *Magai.co*, there will be the need for Google, Meta, TikTok and others to develop tools to weed out bad actors in the AI world.

AI will evolve and mature, and adoption will increase, as will corresponding software to make it safer, and new roles/jobs will emerge that are open to those who can use AI.

At the same time, adoption among the general public will not increase significantly from a pursuit standpoint, but they will be using it unknowingly in many cases in the software they use daily.

Mike Gingerich is an adventurer, triathlete and entrepreneur with a knack for adding value.

87. Sue-Ann Bubacz

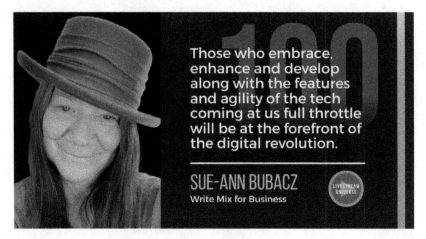

Those who embrace, enhance and develop along with the features and agility of the tech coming at us full throttle will be at the forefront of the digital revolution.

SUE-ANN BUBACZ
Write Mix for Business

I'm sticking with this thought: Technology is about connecting with humans, making things easier for people through automated systems and processes and speeding up things to save the most precious of all resources — your time.

But just like in previous years, I still say the emphasis is on PEOPLE. It's always about people for the win, and ultimately, it's people we do business with, collaborate and exchange ideas with, and generally WANT to interact with as social beings.

Now, to go a little more digital on you, I don't think anyone can predict what the heck is going to happen over the next year or years. I don't think any of us can, truthfully. Not at this never-happened-before pace. The speed of change with the rapid rise of AI and other generative technologies is flying forward so lightning fast that even their makers' heads are spinning, so who am I to guess or pretend to have insight as to where all of this may be heading?

But what I do know is that digital creators and business owners who embrace, enhance and develop along with the features and agility of the tech coming at us full throttle, are the ones who will stay relevant and end up at the forefront in the digital revolution. I also know these new technologies are fun and exciting and offer much to learn and do, allowing you to create in ways never before possible!

I urge you to experiment with and embrace an agile business culture, giving you potential beyond imagination, thrilling creative possibilities and much more, rather than resisting what is already happening with or without you. These changes are poignant to every business, whether digital-first in nature or not. I think we are in for extraordinary times, new frontiers in business and marketing, and impactful changes to our society as a whole as Augmented Reality develops before our eyes, rather than in a future-world dystopia setting.

Technology, good or bad, will continue to be a stronger and more influential force with who we are, integrating more and more ubiquitously into our everyday lives, and well beyond ramifications to just businesses and industries. If we, as content creators and business owners, lead the charge with open minds, curiosity and excitement for the possibilities, who knows what fantastic advances the future of digital holds?

My best advice: Let's go find out!

Now, for your future business. How can you embrace and evolve in this digital whirlwind but still grow your business and better engage with the people you serve? My answer is to build and nurture relationships by creating personally engaging email marketing strategies as part of a concrete content marketing plan.

One of the most overlooked marketing tools, email strategy is often skipped completely or underutilized by businesses in their digital marketing efforts. But to me, this concept is the key. And it has everything to do with how you treat people AFTER the sign-up, opt-in, buy or joining your whatever. It's the after-the-click experience you establish with anyone who says "yes" that matters most of all. How you nurture, interact and connect genuinely to build an ongoing relationship makes all the difference to people who intersect with you.

Ultimately, that difference results in business growth and brand affinity beyond any other marketing method because, firstly, you're invited into someone's inbox. Then, you're given one-to-one access to that person with the chance to embrace and enhance the conversation now open to you. Of course, you need to make the most of this human-to-human opportunity by crafting a valuable, desirable experience so people are excited to hear from you and can't wait to open your emails!

Directly connecting with people by drawing them into your world opens a two-way conversation that's at the center of marketing even in today's digital landscape. Meet people at the center of value — you value them and they value you — for the win!

Sue-Ann Bubacz is an experienced and in-demand content marketing professional, digital content strategist, and writer/editor who works with select small and large businesses and creatives to enhance their communications, elevate influence, and transform creative vision into tangible business results.

88. MARCO NOVO

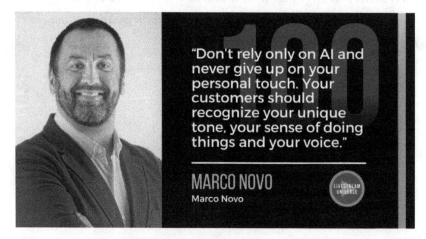

"Don't rely only on AI and never give up on your personal touch. Your customers should recognize your unique tone, your sense of doing things and your voice."

MARCO NOVO
Marco Novo

AI is rising, and we hear about it all the time. Before going to my prediction — I would like to call it a recommendation — I need to make this statement: I'm not an unconditional fan of AI. In my brain, there are mixed feelings, excitement and fear living together.

But let's talk about the good things that AI can bring to us, and I'm going to speak from my experience, with my own business, and from conversations I have with other solopreneurs, small business owners and freelancers.

Some typical traces between us: Lack of time, lack of content ideas, sometimes — if not always — lack of production and editing skills, lack of writing skills, scripting and social media management, and the list could go on and on. Do you relate? Probably you do! These days, you can find tools that could help you with all these challenges.

We can set six different content creation stages: Planning, creation, editing, distribution, evaluation and recycling/repurposing. Just by looking at all these stages, you may get scared, and want to run away from this "content" thing. But, slow down, let me tell you how AI can help you with all of this, stage by stage.

In this stage, you need to know what to create, in which format and when. AI can help you find content ideas, trends and options. Just the fact that AI will remove from you the scary blank sheet syndrome is mind-blowing. Do you need a script? AI can help you. Do you need topics to start getting ideas? AI can help you.

Are you afraid of cameras? AI video generator tools can help you too, even bringing a made-up avatar to speak your brand. It can even write the whole content for you, but I don't recommend that you EVER, by any means, give up on YOUR final personal touch.

Too many filler sounds (ahs and ums) in your audio/video, AI can remove them. It can adjust colors, and lights, so you can look better. It can do spell check for you, awesome help when you're not writing in your native language, like me right now. It can also remove those dead moments on your videos so it can be more energetic.

Where should you post/share your content? When is the best time? AI is there to help you again. It can help you to find the best times to post. It can adjust the posts for each social media platform, helping you with the copy while keeping in mind platform and audience specifications.

Never forget to measure what you're doing, and how that's impacting your business. Here, AI can help you to set which metrics are relevant to your business and how your content creation effort is performing. Sometimes you may get confused or drowned in so much information.

Do you have content that is overperforming? Is there some new social media platform that you want to reach? Do you want to give a new life to some old but evergreen content? AI can do the magic for you. Turn long videos into short vertical ones, blog posts into power points, podcast episodes into blog posts and the list goes on and on.

As you can see, AI can become an amazing help to your content creation journey. Keep in mind that you're using content to improve your business; this will hopefully deliver better results.

AI can be amazingly helpful, flexible and effective, and this is just the beginning. The "I don't have time" excuse is not working anymore. You have almost all the ingredients you need to become successful on your content journey.

I said almost because the "secret sauce" is YOU. You should not rely only on AI. Never give up on your personal touch whether you are a freelancer or a big company. Your customers should recognize your unique tone, your sense of doing things and your voice. An F1 car today is way better than it was three, five, 10 or 20 years ago, but by the end of the race, what makes the difference is the driver! Be the driver!

Marco Novo is a content creator, marketing consultant, livestreamer and remote producer. He hosts "The Special Marcoting Live Show" since 2018 and helps professionals and companies from all over the world to embrace livestreaming as their main content format.

89. JAIME LEGAGNEUR

"Businesses will be looking more toward professional podcast producers for the strategies needed to best use their podcast as a tool to grow their brands — not just their editing skills."

JAIME LEGAGNEUR
Flint Stone Media

It will become easier for new voices and storytellers to join, as more and more barriers to entry continue to come down. And, at the same time, more business owners and brand-builders are learning that knowing the tech and how to put an episode together are only a PART of the equation.

They will be looking more and more toward professional podcast producers for the STRATEGIES needed to best use their podcast as a tool to grow their brands and communities — not just our editing skills.

Producer Jaime ("Jemmy") Legagneur has grown Flint Stone Media since 2014. She is an award-winning podcast host, producer, coach, speaker and industry expert whose team produces more than 60 shows, most notably being featured in Podcast Magazine's "40 Over 40 in Podcasting" list for 2022. She hosts "Podcasting Your Brand" and has founded the Power Moms Network.

90. MARY BARNETT

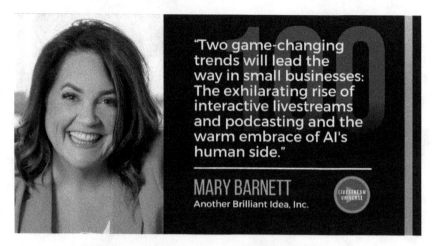

"Two game-changing trends will lead the way in small businesses: The exhilarating rise of interactive livestreams and podcasting and the warm embrace of AI's human side."

MARY BARNETT
Another Brilliant Idea, Inc.

The Era of Dynamic and Engaging Content

As we fast forward to 2024, it's crystal clear that the digital media landscape will be a whirlwind of change. Two game-changing trends are set to lead the way: The exhilarating rise of interactive livestreams and podcasting and the warm embrace of AI's human side in small businesses.

Prediction #1: Unleash the Power of Interactive Livestreams and Podcasts

Podcasts have become a global stage for voices from all walks of life. But with great popularity comes great competition. In 2024, the key to success lies in the speaker-audience connection and the integration of interactive livestreams. This duo creates an experience where the audience isn't just listening — they're actively participating.

As more podcasts enter the scene, creators seek fresh ways to captivate their audiences. Enter interactive livestreams, the bridge between hosts and listeners can use mobile marketing strategies to change the podcasting game with:

1. Real-Time Engagement: Interactive livestreams foster real-time interaction between hosts, guests and listeners. Whether it's a live show or a recording, this instant connection builds a thriving community.

2. Mobile Marketing Magic: Mobile marketing, especially text marketing, turns listeners into VIPs. When a call to action (CTA) is announced on the show, listeners can text in and instantly grab the offer.

3. Automated Text Funnels: After engaging with a CTA, listeners can receive a sequence of text messages that keeps them hooked and informed, bringing them back for new shows and opportunities.

4. Audience Asset Ownership and Business Insurance: Creators need to create and own their audience database. It's an invaluable asset that safeguards against unforeseen changes in the digital landscape. Audiences enjoy immediate rewards, while creators build a direct line to their listeners. The result? Deeper connections, unwavering loyalty and business growth.

Prediction #2: AI Humanizes Small Businesses in 2024

AI is no longer reserved for tech giants; it's a tool for small businesses to shine. It streamlines operations, trims costs and fuels competitiveness.

As we enter 2024, AI's role will expand, offering predictive insights, automating customer interactions and personalizing services. Yet, at the same time, our most valuable asset remains the human connection. Your brand's character, values, and vision set you apart. AI adeptly manages repetitive tasks, but it's your authenticity that gives your business its soul.

With an efficient AI team, you can:

Delegate Mundane Tasks: Let AI manage repetitive chores, from data entry to basic customer queries. This liberates you and your human team for strategic thinking and creativity.

Enhance Customer Engagement: AI-driven chatbots provide instant responses, but true magic resides in human interactions.

Personalize Customer Experiences: AI analyzes data to craft highly personalized interactions so customer preferences help tailor offerings to fortify relationships.

Foster Innovation: With AI handling the mundane, you and your human team can develop new products, explore fresh marketing strategies and stay ahead of the competition.

Let Authenticity Shine: As you delve into AI, remember that your authenticity is your greatest asset. Season your AI team with your uniqueness so it can infuse creativity and stay true to your brand.

Mary Barnett, **known as 'MobileMary' in the industry, founded a boutique marketing firm, Another Brilliant Idea, Inc. in 1988. On her "Brilliant Marketing with Mary" show, she shares tools and tips to make business owners' lives easier!**

By harnessing tools like AI and her software BrilliantMobile.com, she streamlines team efficiency, list building, customer loyalty and profitability!

91. JULIA JORNSAY-SILVERBERG

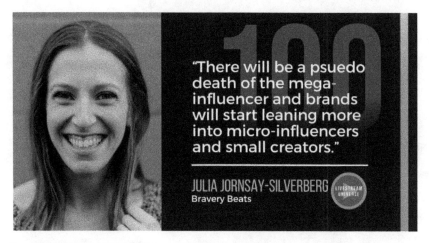

"There will be a psuedo death of the mega-influencer and brands will start leaning more into micro-influencers and small creators."

JULIA JORNSAY-SILVERBERG
Bravery Beats

In 2024, I think that there will be a pseudo death of the mega-influencer and brands will start leaning more into micro-influencers and small creators.

There is so much power that small creators have in building relationships with, and therefore influencing, their audience. I am excited to see more small creators get opportunities to work with bigger brands.

Julia Jornsay-Silverberg is a social media enthusiast who loves to excite and inspire people to cut through the digital clutter by being vulnerable online. As the 2018 recipient of the "Emerging Alumni Award," granted by the University at Buffalo, Julia has 12+ years of digital marketing experience and is known for delivering powerful presentations that leave people feeling empowered.

"Products like XMPie and Chili Publish facilitate the creation of simple business cards or complex real estate brochures in a more user-friendly experience."

JEFF HOWELL
Coverdale Group

Print is not dead, but how you access print will be different. *Variable print technology* has lagged behind. However, technological advancements in addition to forward thinking companies are changing that.

Products like XMPie and Chili Publish are facilitating the ability to create simple business cards or complex real estate brochures in a much more user-friendly experience. Consider this: In the past, creating a custom name graphic that is not generated by a standard font would need a Photoshop expert to manually handle each variable item with that custom element.

Now, with technology like XMPie, integrated with InDesign and Photoshop, it is a matter of building some layers, masks and graphics in Photoshop and the person's name can be a custom graphic, viewed in real time, that would have taken time offline to build in the past.

Chili Publisher, a fantastic web-based variable print tool, has a component where variable graphics with effects can easily be built to create web ads. Where a variable print developer could only think in terms of ink on paper, they can now develop an entire campaign integrating print and digital.

The ability to provide end users with an easy-to-use spec input form and provide a highly-customized item has been on the horizon for many years. Only now, in 2024, has technology advanced enough to make the process much more user-friendly.

Jeff Howell has transitioned his marketing experience and predisposition to authenticity into a successful e-commerce career. His career mission these days is to evangelize the necessity of having a well-executed e-commerce solution integrated into existing marketing strategies while equipping his clients with the knowledge to be successful.

93. DOUG COHEN

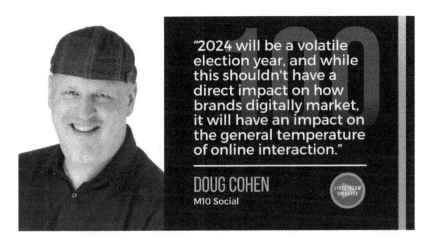

"2024 will be a volatile election year, and while this shouldn't have a direct impact on how brands digitally market, it will have an impact on the general temperature of online interaction."

DOUG COHEN
M10 Social

2024 should shape up to be a pretty consequential year. There is a lot of uncertainty and conflict in the world at the moment — a little more than usual AND it's an election year in the U.S.

So, it will be a volatile year online, and while this shouldn't have a direct impact on how brands digitally market (remember it's never a great idea to allow your brand to get sucked into politics), it will have an impact on the general temperature of interaction online. It will likely get hot, and people will get fatigued.

But this also affords brands an opportunity to be a digital oasis of value/entertainment among the noise. The brands and content creators who will win will be the ones who stay above the fray and make their communities FEEL GOOD. Whether it's livestreaming or podcasting I think people should put their focus on being human.

So how much will AI play into that? I'm not sure. While I don't want to go all "Get Off My Lawn" on AI, I'm also not "all in" on it just yet. I use it sparingly for assistance where it can help me, but I don't lean on it, and that's my recommendation and prediction for how it will be used in 2024 for most of the best marketers.

Maybe it will be more pervasive in 2025 and beyond. Feel free to tell me I'm nuts. Maybe I am. My apologies for not coming in hot with a big bold prediction about some huge shift in the trends for livestreaming, podcasting, social audio, AI and the like, but I feel pretty strongly about my observations just the same!

Doug Cohen runs his digital marketing brand M10 Social out of the Frameable Faces Photography studio he owns with his wife Ally in West Bloomfield, MI. He is an avid blogger, dad to two awesome grown kids, a former Michigan Wolverine football player, history buff, music snob and lead vocalist for Vintage Playboy.

94. JAIME COHEN

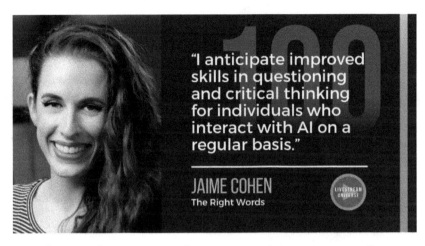

"I anticipate improved skills in questioning and critical thinking for individuals who interact with AI on a regular basis."

JAIME COHEN
The Right Words

AI is shaping the way we work, interact and communicate. Because platforms like ChatGPT, MidJourney, etc. are in their earlier stages, getting answers to our questions means asking the right questions in the right way. It requires specificity.

With this necessity, I anticipate improved skills in questioning and critical thinking developing for individuals who interact with AI on a regular basis!

Jaime Cohen is an internationally-recognized speaker, LinkedIn Learning author and leadership communication coach, who takes you from good to great for more impactful connection and meaningful communication.

95. Mario Fachini

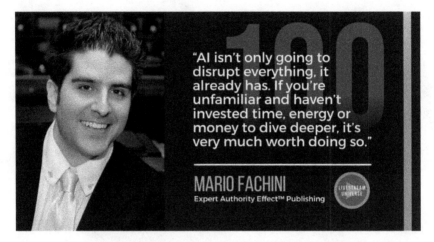

"AI isn't only going to disrupt everything, it already has. If you're unfamiliar and haven't invested time, energy or money to dive deeper, it's very much worth doing so."

MARIO FACHINI
Expert Authority Effect™ Publishing

You are craving connection, and so is your audience. We always have, but the last few years have proven this is truer now more than ever.

Whether podcasting, livestreaming or book publishing, the best tool is the one that works for you; the one that you actually use to get your message to your audience. They need you — and they want more than virtual.

This doesn't mean you need to rent out Caesar's Palace for an event but ask yourself: "What can I do with direct mail?" and just ship it to them.

Additionally, with all due respect, don't make it stupid. A pen or paper clip promoting yourself isn't going to blow anyone's mind. Consider printing out something you already have from your training they can follow along In Real Life vs on their phone — I'd be hard-pressed to find anyone nowadays who doesn't get enough phone time per day.

If you don't have one, create a worksheet that's... "Follow-along-able." If not a course, what about a "5 Day" experience? If not that, promote the heck out of a single "One Off" and have it be the best training of your life... with physical materials.

Here's a freebie marketing pro tip for you:

- Level 1: Minimal or free to join, and they print it off, with basic level access, perhaps no replay.
- Level 2: VIP access and you send them a welcome bundle with physical — at least — required sheet(s). A workbook is better, but publishing your full book is best.
- Level 3: Additional bonus training and/or 1:1 time PLUS all physical and even unannounced bonuses.

There are a million ways to slice this and make it best for you, but no matter what, **make it real and take it offline**.

Additionally, AI isn't only going to disrupt everything, it already has. If you're unfamiliar, barely familiar or know a few buzzwords, but haven't really invested any time, energy or money to "diving deeper," it would be very much worth doing so. It's growing faster than the internet did, and the more you know, the easier it is to learn.

In closing, serve your audience, and you'll win. Give them a great experience and stay on the cutting edge. Here's to your healthy and prosperous 2024.

Business book publishing expert Mario Fachini, *helps successful entrepreneurs — like you — turn your expertise into profits by publishing your own first-class non-fiction business book using The Expert Authority Effect™ Concierge Publishing Method. He is also a two-time #1 best-selling author, Top 100 Entrepreneurship Podcast host, dynamic speaker and enterprise SaaS founder.*

96. PHIL KLUBA

"I'm excited for AI to take over the boring stuff like edit assembly, script structuring and other tedious tasks."

PHIL KLUBA
Press It

Artificial intelligence is here to stay; it will affect all aspects of all businesses. I think it will help creators across livestreaming, podcasting and essentially all aspects of the industry.

I'm excited for it to take over the boring stuff like edit assembly, script structuring and other tedious tasks. AI will help creatives create more effectively.

Phil Kluba is the founder and head producer at Press It, a media production company specializing in livestream, virtual event and XR production. He is also the co-founder of Tall Boy Water, a carbonated canned water brand that provides a fun sober option at any party.

97. Kyle M. Bondo

"OpenAI Whisper will single-handedly destroy the expensive, white-glove transcription service industry."

KYLE M. BONDO
Oncetold

Whispers Through Time

Once upon a time, transcription used to be a luxury that podcasters would ignore. The crafting of converting audio into the written word, while a desired accessibility feature, used to be both a difficult and expensive task.

If you wanted a transcript for your podcast, the easiest way was to rely on your own good old-fashioned human brain power. You, as the podcaster, would simply listen to your own audio and then write down what you heard by hand. This method was accurate but extremely time-intensive.

To solve the time problem of creating a transcript, companies emerged that specialized in transcription services. These companies would (for a small fee) transcribe your audio for you and then return you a finished transcription in a few hours. All you would have to do is upload your audio, go get a cup of coffee and wait. Nice!

Unfortunately, while convenient, the small fee for a minute of audio could add up quickly. Soon, your transcription costs were greater than your hosting costs, making this option expensive to everyone but large studios.

Additionally, a new problem emerged with using these companies. Because labor was cheaper overseas, many of these transcribers would use non-native English speakers. The end result would be transcripts that included strange grammar errors, misspellings and typos. Heaven forbid you had some speaking on your podcast with a Texan accent!

Enter the software developers! They would solve all these problems with code or so they claimed. The early software-driven attempts at transcription automation resulted in mediocre results. Yes, they replaced the strange grammar errors, misspellings and typos the non-native English speakers would make, but in exchange, they created a whole new list of strange grammar errors, misspellings and typos. In some instances, the software was worse than the non-native English speakers and twice as expensive.

Ultimately, a hybrid approach evolved that first allowed a human to correct an automated transcript, then produced tools that would allow you as the podcaster to correct the automation yourself. While useful, this too was time-consuming and came with a price.

Then came OpenAI Whisper. Like most innovations, OpenAI Whisper, a machine learning model for speech recognition and transcription created in 2022, benefited from the previous iterations to solve the speech-to-text problem.

However, when it came to OpenAI Whisper, it could do more than accurately transcribe speech-to-text with minimal errors in both spelling and grammar. It could transcribe English language audio into several other languages and transcribe several non-English languages into English. Additionally, it could do all of this within seconds for pennies.

Overnight, OpenAI Whisper single-handedly destroyed the expensive, white-glove transcription service industry. It was a market disruption on the same scale as the car replacing the horse.

Yet, for podcasters, it was the dawn of a new capability yet to be tapped: Audio Context Search! For years, podcast audio has been a black box; billions of hours of audio locked away in millions of podcast episodes. Without a good episode title, description or even a webpage, podcast audio has been locked away for years.

That was until OpenAI Whisper, and technologies like it, cracked the door open to a treasure trove of forgotten content.

You see, the idea goes something like this: You have a podcast that has NEVER been transcribed. You get access to an OpenAI Whisper tool. You point your OpenAI Whisper tool at ALL of your audio episodes. It accurately transcribes ALL of your audio episodes.

You then point a machine learning (ML) model (which everyone is calling artificial intelligence these days even if it's not) at all of your transcripts to index each and every word. Boom!

A complete "personal" ML-modeled library of your podcast for the price of a cup of coffee!

Now, I want you to think about the possibilities of this concept.

Kyle M. Bondo is an experienced speaker and educator with expertise in the field of podcasting and design. As the CEO and founder of Oncetold, a podcast hosting company, Kyle has extensive knowledge of the podcasting industry and has helped numerous individuals and businesses create successful podcasts.

98. Tamara Thompson

My 2024 prediction is that podcast subscription-based platforms will become more popular for additional episodic content for podcasters, as they provide a way to monetize their podcasts without relying on advertising or sponsorships.

According to a report by Edison Research, 41% of monthly podcast listeners said they would be willing to pay for ad-free versions of their favorite episodes. This also provides a bigger opportunity for podcasters to generate additional revenue.

Tamara Thompson is an investor and founder of Broadcast Your Authority, a content marketing agency for podcasters. Her agency specializes in podcast growth, optimization and distribution, and crafting micro-content sourced from weekly podcasts, strategically positioning companies, influencers and experts as respected industry leaders.

99. VICKI FITCH

"Those interested in how they can train and utilize the technology to serve customers will separate themselves from the average early adopter who sees AI as a quick, time-saving advantage."

VICKI FITCH
Vicki Fitch Enterprises

The virtual landscape as we know it is literally transforming and expanding every moment of every day.

The introduction of new, as well as improved, tools such as *DALL-E 3*, to enhance our AI opportunities, and the low barrier to entry tools like *Aero*, helping creatives to deliver their concepts and ideas in record time, means content creators are producing more content than the average person can consume.

I predict those who are more interested in a deep dive into how they can truly train and utilize the technology to serve customers and clients will easily separate themselves from the average early adopter who sees it as a quick, time-saving advantage.

True creatives will use it as a tool, not a crutch, and will be able to amplify their message because of the quality and precision they are able to achieve with the depth of the content they create.

The seamless blending of AI into the Augmented Reality (AR) world will pop up themed cafes and stores offering immersion adventures on the fly with the possibility of a "write your own ending" addition, where you can modify details to suit your needs before your adventure begins.

Cosplay events will experience a whole new level of intricate costumes and adventure opportunities with like-minded individuals.

I also predict the seedy underbelly will create their own community, that might replace the old Rated R section of the video store with something that startles the senses. It's scary and exciting to watch.

For the average creator, if you love it, you will flourish. If you are in it to make a buck, you will get left behind. Brands, on the other hand, just need to hire the creators who love it in order to watch their vision come to life.

Vicki Fitch is an author, speaker, international business consultant and livestreamer, who is breaking new ground in this area using AI and working to integrate AR into furniture in a division of her company called Furniture Fairy HQ.

100. Doyle Buehler

"To get to somewhere of actual importance in all things digital — and off the Titanic — you need both speed and direction. Pick your topic, players and pace."

DOYLE BUEHLER
Strategystorming.co

Long-term consistency trumps short-term intensity. While we have arrived at the instant-on, always-on, never-off business of digital media in 2023, the busy-ness feeling of the dopamine hit from "firefighting" doesn't last the distance nor get the goals on the board.

In 2024, in order to get to somewhere of actual importance in all things digital, and off the Titanic, you need both speed and direction. Pick your topic, pick your players and pick your pace.

Think of it as the Odyssey of the Entrepreneur. Gather what you can for your journey, be dynamic and think strategically.

You'll drown in other people's content and the flood of AI-generated content bearing down on all of us like a tsunami, otherwise.

Doyle Buehler is on a mission to help businesses go from good to great in the disruptive digital landscape. He believes that by honing our curiosity, creativity and strategic thinking, entrepreneurs can design and create certainty, clarity and a better understanding of their business, creating remarkable value with immense impact and experiences built to last.

101. JANINE NICOLE DENNIS

"Digital platforms and communities of vetted individuals — who commit to a shared purpose and pack a tangible value for users — will likely thrive."

JANINE NICOLE DENNIS
Talent Think Innovations

2024 will mark a moment when social media users start to reexamine and redefine their participation across platforms.

Much of the content we are all seeing online is regressing towards the lowest rung of human expression and in return, many are experiencing some semblance of digital burnout.

Digital platforms/communities that focus on having an inherent value of vetted individuals — who are committed to a shared purpose and pack a tangible value for subscribers and/or users — will likely thrive as a result.

Janine Nicole Dennis is the owner/chief innovations officer for Talent Think Innovations, LLC, a multidisciplinary business strategy and management consulting firm. She is also the author of "The Absurdity of Doing You: Rebel Elegance for the Evolving Soul" and the host of the "Growth on my Terms" podcast.

102. TIMOTHY KIMO BRIEN

"In the United States, politicians will see an opportunity to force bigger podcasts to register with the FCC or some other agency."

TIMOTHY KIMO BRIEN
Create Art Podcast

In the United States, politicians will see an opportunity that Canada has already taken, to force bigger podcasts to register with the FCC (Federal Communications Commission) or some other agency. This will be supported by more traditional forms of media that already are subject to FCC regulations.

The reason for this will be the claims of Joe Rogan and Ben Shapiro and other big podcasters that interfere with or directly contradict government plans or programs such as COVID-19 vaccination or UFO revelations.

Tmothy Kimo Brien produces podcasts for the US Federal Government as well as his own. He has been making podcasts since 2006.

103. ROB CAIRNS

'The biggest change is more and more podcasters doing video for their podcasts. As YouTube makes podcasts more of a priority, video will become more important."

ROB CAIRNS
Stunning Digital Marketing

A lot has changed in podcasting in the last few years. Equipment is better now and cheaper than ever before. AI is impacting podcasting in a big way in everything from making show notes to podcast production. We are truly in interesting times.

Despite all these changes, I believe the biggest change is more and more podcasters doing video for their podcasts. As YouTube makes podcasts more of a priority, video will become more important. I have seen a direct increase in the number of people enjoying my podcast by doing video.

Rob Cairns is the founder, CEO and chief creator of amazing ideas at Stunning Digital Marketing. He is the creator of the SDM Show Podcast.

104. NICOLE CHRISTINA

"A podcast interview that brings out spontaneity and vulnerability in the guest will become deeply appreciated and sought after by audiences."

NICOLE CHRISTINA
Zestful Aging Podcast

The more AI permeates our world, the more people will crave real, genuine human connection.

AI simply can't offer us a true human experience, and our mammal brains are wired to need this. Audiences want to feel a sense of belonging and a feeling of being understood.

My prediction is that a podcast interview that brings out spontaneity and vulnerability in the guest will become deeply appreciated and sought after by audiences.

Nicole Christina hosts the Zestful Aging Podcast, a multi-award-winning interview show heard in 106 countries. Her guests are change-makers from a variety of disciplines. She calls herself "the Terry Gross of aging well." The podcast is also streamed on WCNY-FM in Syracuse, New York.

105. ZACK BORDEAUX

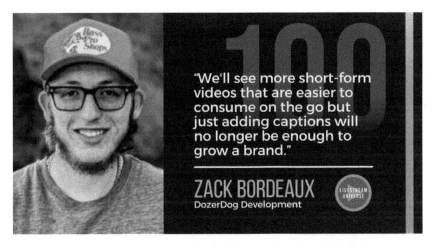

"We'll see more short-form videos that are easier to consume on the go but just adding captions will no longer be enough to grow a brand."

ZACK BORDEAUX
DozerDog Development

I don't think YouTube is going anywhere but I do think we will see a change in video styles. We'll see more short-form videos that are easier to consume on-the-go but just adding captions will no longer be enough to grow a brand.

There is going to be an even bigger divide between the best and the rest — and those who take the time to create high-quality content that engages with their viewers will be rewarded.

Zack Bordeaux is the serial entrepreneur, content creator and thought leader who delivers the energy, the effectiveness and the embodiment of carving out your own path for ambitious disruptors and a well-honed work ethic who will not stand for living a life of mediocrity and mindlessness and are ready to pursue potential and 10x their tenacity.

106. Christine Gritmon

"2024 is really going to be about honing smaller conversations with more curated communities."

CHRISTINE GRITMON
Christine Gritmon Inc.

For me (and many others), 2023 was about Twitter's rapid demise. Not sure about assigning cause-versus-effect status to this one, but I feel this marked the end of the era of the one-to-many open conversations, discovery and information-sharing that defined social media for so long.

Add in AI and its capacity for spreading misinformation, and I think 2024 is really going to be about honing smaller conversations with more curated communities.

I will miss being able to hop on Twitter and discover new people, interesting conversations and information about current events — it was fun while it lasted — but now I'll have to seek out more deliberate communities with whom to have more closed conversations.

Christine Gritmon empowers professionals to step into their personal brands in a bigger, bolder way on social media. You CAN do it — she'll teach you how! She's spoken on stages worldwide and is a frequent expert guest on podcasts, livestreams, Twitter chats and blog posts, as well as hosting her own weekly podcast, "Let's Talk About Brand."

107. TIM GILLETTE

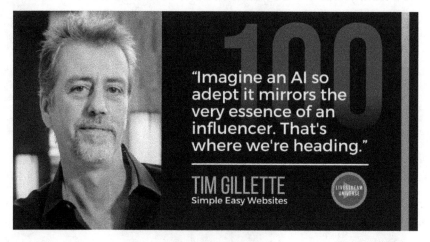

AI Personalization

Imagine an AI so adept it mirrors the very essence of an influencer. That's where we're heading.

As content creators increasingly pivot to AI tools, the real game-changers will be those who don't just use AI but merge with it.

They'll train these tools to emulate them so flawlessly that viewers will be left wondering: "Is this the influencer or the AI?" The true mark of success will be that indecipherability.

Tim Gillette, life-long entrepreneur and innovator, hosted the renowned Blog and Video Con in Dallas (2018-2021). He's an international speaker and best-selling author of titles like "You Can Always Get What You Want." The visionary behind Simple Easy Websites, Tim embodies the spirit of innovation, passion and relentless pursuit of growth.

108. ANITA SONYA

"We can have all the creativity we want in easy ways from voice AI integration to audience engagement metrics. More software and apps will provide this."

ANITA SONYA
LivePodcast Media

AI has evolved and changed everything in our work and business and the way we live. It's not only about creating content, coding and collecting data anymore.

The next question is more about how to create businesses or jobs and grow with it, especially since there will be more GPT builders from voice and photo creation, and there will be more video creation.

It's amazing how it can help you with marketing, team communication and AI as a service (AIaaS). It's true that if you don't get along with it, your worries will come true that you can lose your job or business or be left behind since the others have already run with it, breaking the barriers and across languages.

Get ready to overcome the challenges in AI content generation and copyright, privacy concerns, regulatory changes, algorithm changes and dependence, content overload, ethical use of AI itself, and accessibility and inclusivity.

Meanwhile, we can have all the creativity we want in easy ways from voice AI integration to audience engagement metrics. More software and apps will provide this. More livestreaming and podcasting platforms will facilitate you with enhanced interactive features, automated content creation, language translations and monetization opportunities. So, no need to worry because AI can support remote work too.

It's very exciting to see how fast we face the changes because of AI every day, don't you think? Just like always, let's see the positive and bright side so we can keep going.

Anita Sonya is the founder of LivePodcast Media including LivePodcast Insider helping business professionals and content creators transform their businesses through livestreaming. She organizes multiple international events supported by leading brands such as Blubrry, Ecamm Network, Wave.video and Restream, bringing together top expert speakers from diverse industries.

109. Carlos Phoenix Jimenez

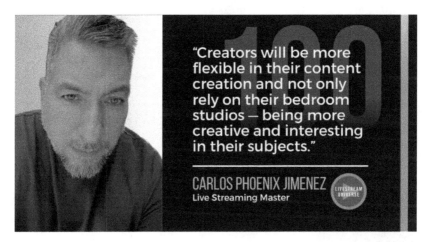

"Creators will be more flexible in their content creation and not only rely on their bedroom studios — being more creative and interesting in their subjects."

CARLOS PHOENIX JIMENEZ
Live Streaming Master

I foresee a future where the distinctions between livestreaming, entertainment, podcasting and YouTube blur. Podcast networks may seamlessly integrate into streaming apps, TV downloads and various applications, empowering them to enhance brand control and explore diverse monetization avenues.

Hollywood will also join the fun and create their own podcast networks and have them side by side with their films and tv series.

Creators will start to be more flexible in their content creation and not only rely on their bedroom studios, being more creative and interesting in their subjects.

The future is bright and Covid has helped it move forward.

Carlos Phoenix Jimenez, a pioneering Live Streaming Master CEO and accomplished entrepreneur, transitioned from an early career in Sci-Fi/Fantasy illustration to making a mark in Hollywood.

With a background in advertising and film production, Carlos now focuses on the tech for the future of content and distribution.

110. STEVEN HEALEY

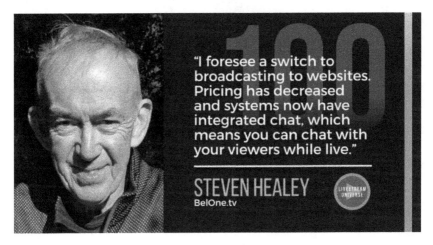

"I foresee a switch to broadcasting to websites. Pricing has decreased and systems now have integrated chat, which means you can chat with your viewers while live."

STEVEN HEALEY
BelOne.tv

I have been live broadcasting on social media channels like Facebook, YouTube and Linkedin for eight years. In the following year, I foresee a switch to broadcasting to websites.

Two factors will drive this change: Pricing has decreased, and systems now have integrated chat, which means you can chat with your viewers while live.

Live video drives traffic to your website, where viewers will watch pre-recorded videos and visit your online shop, which will become standard practice.

On your website, you are in charge, and your website becomes part of your marketing mix.

Steven Healey has the honor of working with some of the most innovative and exciting companies in the live broadcasting industry.

He is a content creator and ambassador for BeLive and a streaming partner with OneStream.Live

111. GORD ISMAN

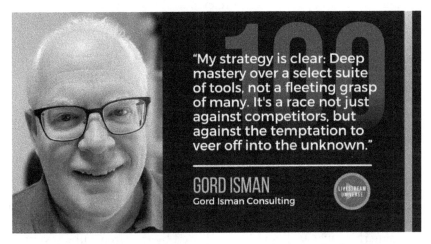

"My strategy is clear: Deep mastery over a select suite of tools, not a fleeting grasp of many. It's a race not just against competitors, but against the temptation to veer off into the unknown."

GORD ISMAN
Gord Isman Consulting

Navigating the AI Gold Rush: My 2024 Formula for Success

Gripping the steering wheel, I feel the thrum of my Formula 1 racecar's engine, a pulse echoing my own. The track ahead is a whirlwind of 2024's digital media innovations; AI tools flashing past like rival racers vying for the lead. Each tool tempts with the promise of new speed, new efficiency.

Yet, I hold my course, focused.

I'm not lured by the shimmer of every passing innovation; I've learned the power of knowing my tools as intimately as the contours of my car's dashboard.

The race of 2024 unfolds at a frenetic pace.

Around me, content creators weave in and out, their eyes darting to the latest AI advancements, their hands reaching for each new tool in a dance of uncertainty.

But there's a different rhythm to my approach.

I maneuver with a select few AI tools, each turn on the track a testament to our synergy. They're extensions of my vision, gears and circuits melding seamlessly with my creative impulses.

As each curve of the track brings new challenges, my strategy is clear: Deep mastery over a select suite of tools, not a fleeting grasp of many.

It's a race not just against competitors, but against the temptation to veer off into the unknown.

The real question as we surge through 2024 — a question I ponder as my car slices through the digital whirlwind — is how will you navigate this race? Will you be captivated by the siren call of every new tool, or will you steer a path of focused mastery, your chosen AI technologies an extension of your own racing finesse?

Gord Isman is a video strategist, video editor and YouTube consultant. He helps entrepreneurs, coaches and small businesses utilize video to get their message out to the world.

112. IAN ANDERSON GRAY

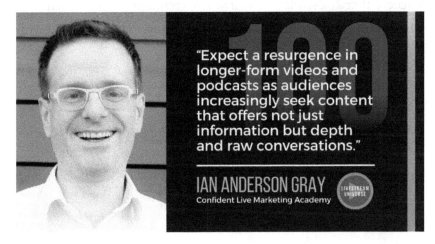

"Expect a resurgence in longer-form videos and podcasts as audiences increasingly seek content that offers not just information but depth and raw conversations."

IAN ANDERSON GRAY
Confident Live Marketing Academy

As we step into 2024, we're on the cusp of an unprecedented leap in digital media driven by the exponential growth of artificial intelligence.

ChatGPT 5 will mark a significant evolution from its previous versions and will unlock new frontiers, improving the quality of AI-driven content repurposing and editing. But this technological advancement will underscore the ever-growing importance of live video.

In a landscape teeming with AI-generated content, live video promotes a medium of authenticity, offering a uniquely human touch that resonates deeply with audiences.

We've seen the meteoric rise of short-form, vertical-format videos.

In 2024, I see the pendulum swinging back towards a desire for depth and connection. Expect a resurgence in longer-form videos and podcasts as audiences increasingly seek content that offers not just information but depth and raw conversations.

I believe that 2024 will be a year where technology and human connection converge in digital media, creating a landscape rich with both innovation and heartfelt storytelling.

Ian Anderson Gray is the founder of the Confident Live Marketing Academy and is the host of the Confident Live Marketing Podcast. He helps entrepreneurs to level up their impact, authority and profits by using live video confidently. He's an international speaker, trainer, teacher and consultant. He trained as a professional classical singer and lives near Manchester in the UK.

113. STEPHANIE GARCIA

"Whether parent or foodie, DIY-er or aspirational influencer, every family member will livestream and generate user content casually and creatively."

STEPHANIE GARCIA
Lights, Camera, Live

In 2024, everyone is a content creator. Creating videos and podcasts will become as ubiquitous as baking a cake, as the average household transforms to accommodate this new digital role.

Expect cluttered corners replaced by outfitted garage studios prime for family vlogging, and teen bedrooms outfitted with ring lights and green screens instead of band posters. The ambition to entertain and influence need not look further than one's own home.

So, whether parent or foodie, DIY-er or aspirational influencer, every family member will livestream and generate user content casually and creatively. Get ready: Every house will become a digital broadcast hub in 2024!

Stephanie Garcia is a live video strategist for brands and entrepreneurs — from moonwalkers to master chefs — who want to go from unknown to unforgettable.

A digital marketing expert with 15 years of ad agency experience, she has helped thousands of her clients and students get on the Fascination Fast Track™ to ignite their ideas and become confident on camera.

2024 Contributors

PART III
RESOURCES

NEXT STEPS

Thank you for choosing this book.

Stay on top of the latest trends and tactics for creating content that serves your audience and builds your business. It's free and easy to benefit from the following Livestream Universe offerings:

1. **Get the *StreamLeader Report* monthly email**: Our free report delivers livestreaming and digital media info and strategies to keep you ahead of the competition.

   livestreamuniverse.com/streamleader

2. **Subscribe to the YouTube channel**: Catch our new videos as they drop, including tutorials, reviews and interviews with leading creators and entrepreneurs.

   youtube.com/c/livestreamuniverse

3. **Join Our Facebook Community**: Continue the conversation with the book's author and contributors in the Livestream Universe Community Facebook Group.

   facebook.com/groups/livestreamuniversecommunity

WHAT'S IN ROSS' CREATOR TOOLKIT?

NEW ADDITIONS FOR VOLUME 4

Generative AI

Chat GPT: "ChatGPT is an advanced AI language model developed by OpenAI, capable of understanding and generating human-like text based on the input it receives," said ChatGPT when I asked it for a one-sentence response to the question, "What is ChatGPT?"

livestreamuniverse.com/chatgpt

Short Videos

OpusClip: The demand for content for Instagram and Facebook Reels, YouTube Shorts and TikTok often exceeds our bandwidth to create and produce short-form vertical videos from scratch. OpusClip speeds your time-to-market by taking a talking-head video recording and repurposing it with one-click into 10+ short clips for sharing on social media with captions, emojis, highlighted keywords and b-roll.

livestreamuniverse.com/opusclip

Video Hosting & Storage

Streamable: Hosting or storing your videos online has typically been an expensive proposition. No longer. Streamable makes it easy and affordable to upload and share your videos, including a clean, branding-free player to embed on your own website.

livestreamuniverse.com/streamable

Low Profile Boom Arm

Elgato Wave Mic Arm LP: If you want a less cluttered look for livestreams and recorded video, but still want to get top-quality audio for your podcast version, a low profile boom arm is a great choice for meeting both goals. It also has a more modern appearance than the traditional overhead boom arms.

livestreamuniverse.com/elgato-LP

Podcast Media Host

Spreaker: I remain thrilled with Captivate as the podcast media host for most of my podcasts and highly recommend it. For the StreamLeader Report, we are using Spreaker, an old favorite that has some great features. Spreaker's website and listening app offer discovery and the podcast player looks great on your website.

livestreamuniverse.com/spreaker

Books

Live Streaming Made Easy: A Step-by-Step Guide to Going Live: If you're brand new to livestreaming or need a refresher, you won't find a better quick-start guide. Kevin Kolbe provides everything you need to know for your first or next livestream. Read it tonight and go live tomorrow.

livestreamuniverse.com/kolbe-book

Leverage Livestreaming to Build Your Brand: Start, Master, and Monetize Live Video: Some of the tools and tech may have changed since Jennifer Quinn aka JennyQ wrote this comprehensive guide to livestreaming, but the core strategies and approaches to live video remain relevant for anyone leveraging the medium to achieve marketing and business success.

livestreamuniverse.com/jennyq-book

The Amazon Self Publisher: How to Sell More Books on Amazon: If you want the game plan I used for self-publishing the #100Predictions series, you'll find much of it in this three-book series by Dale Roberts covering keywords, marketing and promotion, and Amazon reviews.

livestreamuniverse.com/dale-book

LIVESTREAMING & VIDEO PRODUCTION SOFTWARE

Cloud-Based Livestreaming Software

StreamYard: Exceptional stability unmatched by any other browser-based livestreaming studio. Improved quality for editing and podcasting thanks to the addition of local recording with separate audio and video tracks for each participant. New "StreamYard On-Air" webinar platform is a winner.

livestreamuniverse.com/streamyard

Livestreaming Software for Mac

Ecamm Live: Release your inner creative and build your scenes out the way you always wanted. Lots of user control without the learning curve required by Wirecast or OBS. The team provides helpful content for learning different ways to use Ecamm Live.

livestreamuniverse.com/ecamm

Mobile Livestreaming Software for iOS

Switcher Studio: Remains the industry standard for iOS live video production. Use your iPhones and iPads for multi-cam video, insert titles and graphics, edit live and stream anywhere. New features to support live shopping.

livestreamuniverse.com/switcherstudio

Professional Livestreaming Software

Wirecast: If you want to produce a TV-quality livestream without breaking the bank, Wirecast is a robust, professional livestreaming solution that gives you complete control.

livestreamuniverse.com/wirecast

Video Editing & Screencasting

ScreenFlow: While I continue to experiment with text-based editing, ScreenFlow remains my trusted tool for video editing and screencasting. More advanced than iMovie and easier to use than Premiere Pro or Final Cut Pro.

livestreamuniverse.com/screenflow

Descript: This text-based video and audio editing software is a revolutionary advancement that both simplifies the editing process and contains powerful features for upgrading the quality of your audio and presentation of your video. It transcribes your video or podcast automatically upon upload. Whether as your sole editing tool or in combination with a traditional video editor, this is a game changer.

livestreamUniverse.com/descript

YouTube Optimization

TubeBuddy: If you don't have the free Chrome browser extension, follow the link below and download it now. Paid plans give you the full benefit of this essential tool for growing your YouTube channel, researching topics and getting more eyeballs to your videos.

livestreamuniverse.com/tubebuddy

Video Hosting & Storage

Streamable: See "New Additions for Volume 4"

livestreamuniverse.com/streamable

Video Monetization

Uscreen: It's time to monetize livestreaming and on-demand video. Uscreen makes it easy to sell subscriptions, courses and membership communities. Run your own Netflix-style streaming service.

livestreamuniverse.com/uscreen

Video Overlays & Backgrounds

Streamer Templates: Stand out on live video with customizable templates designed specifically to work with StreamYard's layouts.

livestreamuniverse.com/templates

PODCAST SOFTWARE

Podcast Media Host

Captivate.fm: Designed with the podcaster in mind, it's an unmatched user experience with all the tools you need to manage your podcast or network. It's my favorite podcast hosting service.

livestreamuniverse.com/captivate

Spreaker: See "New Additions for Volume 4"

livestreamuniverse.com/spreaker

Podcast Recording & Remote Interviews

Squadcast: Record audio & video interviews in studio quality from anywhere, with anyone, around the world. Create engaging audio and video content with an intuitive platform that allows remote recording and simplified editing.

livestreamuniverse.com/squadcast

Podcast Website

Podpage: Build a beautiful podcast website in five minutes from your RSS feed. Lots of customization options to build a multi-purpose fully functional website like the one I have at RossBrandRecordings.com. Unbelievably affordable premium plans include hosting and custom domains.

livestreamuniverse.com/podpage

AI TOOLS

Transcripts & Audio/Video Editing

Otter.ai: AI transcripts of audio and video content.
livestreamuniverse.com/otter

Descript: AI transcripts and audio/video editing. See "Video Editing & Screencasting" above for more.

livestreamUniverse.com/descript

OpusClip: See "New Additions for Volume 4"

livestreamuniverse.com/opusclip

Copywriting

Jasper: Use the help of AI to write high converting copy for increased conversions and higher ROI. Templates for everything from blogs to video scripts.

livestreamuniverse.com/jasper

Multipurpose

ChatGPT: See "New Additions for Volume 4"

livestreamuniverse.com/chatgpt

AUDIO & VIDEO GEAR

Links to the video & audio gear discussed below:

kit.co/rossbrand/video-gear-recommendations

kit.com/rossbrand/audio-gear-recommendations

Studio Microphones (XLR)

Shure SM7B: While it rocketed to popularity in recent years among podcasters and streamers, it's been a radio and music industry staple for decades. It's the mic I use in my home studio. I love its versatility and warmth.

Electro-voice RE20: Though not as popular in the digital media space as the SM7B, this was my favorite mic when working in radio. It deserves far more attention than it gets from podcasters and live video hosts. The new black edition looks good on video too.

Heil PR40: Not the choice for every voice, but on some voices this professional broadcast mic sounds amazing. Not nearly as gain hungry or as expensive as the mics listed above.

USB Microphones

Samson Q2U: If you're new to livestreaming and podcasting and don't have mixers and audio interfaces sitting around your home, then there is no reason to start with any other mic. It sounds far too good for the price and nobody I've recommended it to has ever complained.

Shure MV7: If you want the option of plugging in USB but need to sound like you're using one of the better studio mics, then this hybrid is an excellent choice. Not sure there is a better sounding dynamic USB option.

Elgato Wave 3: Some people want to keep the mic out of their shot or at least not as close as most of the options above require. Although it goes against my principles to recommend a condenser mic for home studio work, this is a strong choice.

Remote Interviews & Field Reporting (XLR)

Electro-voice RE50: When I need to go live from events or capture interviews in noisy environments, this is my top choice and I really don't have a runner up. There's a good reason — it's the broadcast industry standard microphone for electronic newsgathering.

All-Purpose Microphone (XLR)

Sennheiser e835: Want one affordable, professional mic that can be used in the studio, out in the field and on stage? This remarkable handheld dynamic mic travels with me whenever I'm on the road, usually right next to the RE50.

Audio Interfaces, Mixers and Recorders

Sound Devices MixPre-3: It's got the best sounding preamps of any device in its class and its analog-to-digital converters are excellent as well. The MixPre makes a wonderful audio hub in your home studio and an exceptional field recorder for video shoots. There's a learning curve to using it, but the reward is professional-quality audio.

Rode Rodecaster Pro 2: This soundboard-like device just looks like a lot of fun to use. It's become the go-to choice for many podcasting studios. If you are producing while livestreaming or recording, this mixer/interface/recorder is a good blend of quality and ease of use.

Zoom Podtrak P4: Not as cool as the Rodecaster Pro, but lightweight and portable with high-quality audio at a fraction of the price.

Focusrite Vocaster: Both new and experienced podcasters and livestreamers will love the ease of use and smart feature set included in the Vocaster interfaces from Focusrite. More than enough gain to power an SM7B or RE20, it also has auto gain, vocal presets and can send audio into your camera.

Headphones

Sony MDR-7506: I use these every day. They're comfortable, work great for editing and are one of the better aesthetic choices if you wear headphones on video.

Standard Microphone Boom Arms

Rode PSA-1 Boom Arm: Perfect for heavier studio mics like the SM7B, RE20 and PR40.

Neewer Mic Boom Arm: Works well with handheld mics such as the Q2U and e835 at a much lower price than the Rode.

Low Profile Microphone Boom Arm

Elgato Wave Mic Arm LP: See "New Additions for Volume 4"

Webcams

Logitech c922x (or c920): These webcams are all you need to go live or record with good quality video.

Mirrorless Camera

Sony A6400: When it comes to mirrorless video cameras, especially for livestreaming, Sony is a winner. Many worthy options from Sony, but I went with this one and couldn't be happier with it.

Green Screen

Elgato Green Screen: "What green screen do you use?" That's one of my most frequently asked questions over the past few years. I don't have enough data to compare it to others on the market, but I can say that it came highly recommended by photographer and video creator Mike Murphy (contributor to Volume 2), it's well-made and I have no complaints. It's also easy to setup and take down in your home studio.

GRAPHICS & SHORT VIDEOS

Streamer Templates: Stand out on live video with customizable templates designed specifically to work with StreamYard's layouts.

livestreamuniverse.com/templates

Canva: Popular DIY graphic design platform that does so much more than provide templates for static social posts. Edit short-form videos, design your website, schedule your social posts and write out articles with AI. The background removal tool is worth the price of the Pro Plan by itself.

livestreamuniverse.com/canva

PlaceIt: Create mockups, logos, social posts and videos in seconds. Great for intro videos for livestreams.

livestreamuniverse.com/placeit-2

OpusClip: See "New Additions for Volume 4"

livestreamuniverse.com/opusclip

Wave.video: Create thumbnails, backgrounds, overlays, intros, outros and other image and video assets to use in your content.

livestreamuniverse.com/wave

Depositphotos: High-quality stock photos and video b-roll.

livestreamuniverse.com/depositphotos

Creative Fabrica: Fonts and design elements for your online graphics and physical products.

livestreamuniverse.com/creative-fabrica

WEBSITE, EMAIL & MARKETING

WordPress Hosting

Rocket.Net: Kim Doyal recommended Rocket.Net when I was searching for a new webhost. I'm absolutely blown away by how blazing fast it is since changing hosts. Rocket.Net completed the move for me at no extra charge within an hour from when I signed up.

livestreamuniverse.com/rocketnet

WordPress Themes & Page Builders

Elementor Builder: Build and customize every part of your website intuitively with Elementor's visual builder. Zero coding required.

livestreamuniverse.com/elementor

Elegant Themes: Divi is the most popular premium WordPress theme in the world and comes with Divi Builder.

livestreamuniverse.com/divi

Podcast Website

Podpage: Build a beautiful podcast website in five minutes from your RSS feed. Lots of customization options to build a multi-purpose fully functional website like the one I have at RossBrandRecordings.com. Unbelievably affordable plans include hosting and custom domains.

livestreamuniverse.com/podpage

Domains

NameCheap: Tends to have lower prices than its competitors along with a beautiful website, convenient mobile app and easy-to-use interface.

livestreamuniverse.com/namecheap

Email Marketing

ActiveCampaign: Email list building on the world's leading email automation platform.

livestreamuniverse.com/active

Course Delivery & Sales

Thinkific: Popular choice for building online courses and membership communities.

livestreamuniverse.com/thinkific

Social Media Management

Agorapulse: The premier tool for social media scheduling, listening, engaging and analytics just got even better with the ability to schedule TikTok videos right from inside Agorapulse. Great for teams and collaboration. Indispensable when managing many accounts. Customer service is top-notch.

livestreamuniverse.com/agorapulse

GLOSSARY

Special thanks to OpenAI's ChatGPT and Google's Bard for assistance with definitions and explanations provided in the glossary.

Aero: An augmented reality (AR) content creation tool that allows users to easily design, build and share interactive experiences in AR, bringing digital elements into the real world through a user-friendly interface.

American television: In this book, it refers to linear television broadcast in the United States as opposed to video distributed online. However, in other contexts, the specific formulation, "American television," could be a reference to the movie "Bananas," as the phrase was uttered in bombastic fashion by Howard Cosell (playing himself) while covering a fictitious foreign assassination in the manner of a championship boxing match.

AVOD: (Advertising-Based Video On Demand) is a digital streaming service model where users can access content for free and the revenue is generated through advertisements.

Bitcoin blockchain: A decentralized digital ledger that records all transactions across a network of computers, ensuring the security and integrity of Bitcoin transactions through cryptographic principles.

Bitcoin Ordinals: A type of digital artifact inscribed directly onto individual Satoshis, the smallest unit of Bitcoin, allowing for the creation and tracking of unique, non-fungible tokens within the *Bitcoin blockchain*.

Charlie: An AI agent that assists companies in achieving enhanced business results through features like advanced natural language processing, content generation and automated data analysis.

Claude: An AI chatbot and large language model created by Anthropic AI, emphasizing ethical AI development. It is designed to perform tasks such as summarization, editing, Q&A and decision-making with a focus on safe and responsible AI practices

CTV: (Connected TV) refers to any television set that can connect to the internet and access content beyond what is available via the normal offering of cable or satellite providers.

DALL-E 3: An advanced AI image generation tool developed by OpenAI that creates highly detailed and creative images from textual descriptions, offering an intuitive and powerful way for users to bring their visual ideas to life.

Dark Web: A hidden part of the internet that is intentionally concealed from standard search engines and often associated with anonymity, illicit activities, and encrypted communication.

Death scrolling: The habit of continuously and compulsively scrolling through distressing or negative news and social media content, often to the point of feeling overwhelmed by the volume of distressing information.

Deepfake: A type of manipulated media created using artificial intelligence that convincingly replaces a person's likeness in images or videos with another person's, often raising ethical and privacy concerns.

DemandGen: Short for Demand Generation, DemandGen is a marketing strategy focused on creating awareness and interest in a company's products or services with the goal of building a long-term consumer pipeline.

Descript: A multimedia editing software that offers comprehensive tools for both audio and video editing, enabling creators to record, produce and repurpose podcasts and video content using innovative functions like text-based editing and automatic transcription.

Dolby Atmos: An advanced audio technology that creates a three-dimensional sound experience, allowing sound to move freely around the listener, including from overhead, to provide a more immersive and lifelike audio experience in movies, music, and games.

insoundz: Generative AI audio enhancement solution takes streamed or pre-recorded audio and video, then applies generative AI signal processing to output the same streams with greatly increased studio-like audio sound, keeping only sound you want, removing everything else.

Language Learning Model (LLM): A type of artificial intelligence (AI) system designed to understand, interpret and generate human language based on vast amounts of text data, often used for tasks like translation, content creation and conversation.

LED volume stages: Sophisticated soundstage environments equipped with LED panels that create dynamic virtual backgrounds and settings, enabling creators to produce high-quality content with immersive visual effects and realistic virtual environments, a trend that is gaining traction among serious content creators nationwide.

Magai: A platform that provides an array of AI-driven tools for content creators, featuring real-time data and insights to aid in content optimization and strategy enhancement.

MidJourney: An AI-powered tool that generates creative and complex images based on textual descriptions, enabling users to create unique and artistic visuals through an intuitive interface.

Nomono: Nomono Sound is a small, portable podcast studio that records and syncs up to four voices to Nomono Cloud for enhancement, editing and collaboration.

Nostr: An acronym for "Notes and Other Stuff Transmitted by Relays," Nostr is a decentralized social media protocol designed to give users control over their data and resist censorship, allowing for a range of applications from social networking to messaging and games, with cryptographically validated posts.

Podcasting 2.0: An evolving set of upgrades bringing interactivity and richer features to podcasts, like chapters, transcripts and bonus content, aiming to make listening more personalized and engaging. Podcasting 2.0 offers new monetization channels beyond traditional ads, like direct listener micropayments, NFTs linked to episodes and interactive features enhancing brand partnerships. It's about building deeper listener engagement to unlock diverse and innovative revenue streams.

Quantum computing: A type of computing that uses quantum-mechanical phenomena, such as superposition and entanglement, to perform operations on data, enabling it to solve complex problems much faster than traditional computers. Quantum computing is still in the early stages of development and primarily exists in research and experimental settings. Quantum computers that can outperform classical computers for practical, everyday applications are not yet widely available.

Quantum-powered AI: The application of quantum computing technologies to enhance artificial intelligence systems, potentially allowing them to process information and learn at extraordinarily faster rates compared to AI running on classical computers. Quantum-powered AI is largely theoretical and experimental at this stage.

RSS: In podcasting, RSS (Real Simple Syndication) is a vital distribution tool that allows podcasters to automatically deliver new episodes to listeners across various platforms, ensuring uninterrupted access to content, even if faced with restrictions or removal from a specific service, thereby maintaining broad audience reach and control over their content.

Satoshi: The smallest unit of a Bitcoin, like a "cent" to a dollar. There are 100 million Satoshis in one Bitcoin, making it handy for tiny transactions. (Think "pocket change" for the digital age).

SVOD: (Subscription Video On Demand) is a digital streaming service model where users pay a recurring fee to access a library of content without advertisements.

Variable print technology: A form of digital printing that allows for the customization of individual pieces within a larger batch, enabling personalized text, graphics and images on each printed item.

ACKNOWLEDGEMENT

There are 113 people I want to recognize right off the bat without whom this book wouldn't be possible. Their names are listed in the Table of Contents and at the heading of each Predictions chapter. They are, of course, the contributors whose insights and forecasts fill the heart of this book. I am grateful for your willingness to share your professional expertise with our readers.

Danielle Camorlinga designed the book cover for the fourth straight volume. How talented is she? From day one, her design skills have made this series standout on Amazon, across the web and on bookshelves. I am extremely grateful, Danielle, for your irreplaceable work.

Thank you to Rob Greenlee, Claudia Santiago and Rebecca Gunter for writing the foreword to Volume 4. My favorite hour of the week is when we're having a blast co-hosting the StreamLeader Report Live Panel. You are all much more than broadcast partners; you are loyal and supportive friends.

Thanks to Kevin Kolbe for asking me to write the foreword for his first book, *Live Streaming Made Easy* (read my entry on the book in "New Resources for Volume 4"). It's an honor to be a part of your excellent guide to livestreaming. Kevin has also generously promoted *100 Predictions* while marketing his book.

I also want to thank Mae Maralit for helping with graphics and being a sounding board for content ideas. Thanks for all you do — the book is better because of your input.

Thanks to self-publishing legend and good friend Dale Roberts for keeping me informed of the latest changes in the industry. I've never gone wrong following his game plan for Amazon KPD.

Thanks to Jennifer Quinn "JennyQ," Mitch Jackson and Jennie Mustafa-Julock for being a part of this annual project every year since it started as a blog post in 2016. Talk about showing up!

I'm thankful for three people who were part of the first Livestream Universe Predictions nine years ago — and contributed to other blog editions — returning to make their first appearances in the book series. Welcome back to Janine Nicole Dennis (née Truit), Karen Graves and Jonathan Tripp!

Thanks to Beauty Bubble for seeing value in this project back in the early years and always encouraging me to keep it going. I greatly value your enthusiasm and advice.

Thanks to the team at Podfest Expo (Chris Krimitsos, Andrew Weiss, Wendy Brandt, Larry Roberts and Lee Uehara) for supporting authors with their Author Meet & Greet at the annual conference in Orlando and providing mentions at Podfest local meetups, PodLab sessions and in the Podfest newsletter.

And last but not least, thanks to my parents for always being my rock. I love you.

ABOUT THE AUTHOR

Ross **Brand** is the connection between brands, early-stage startups and customers through the power of live video. He's served as an on-air personality, brand ambassador and consultant for industry brands such as StreamYard, BeLive, LiveU and Wirecast. Ross was instrumental in the rapid growth and adoption of StreamYard through his shows and network of digital influencers. Switcher Studio selected him as one of "5 Live Video Experts to Follow" and Klout named him the world's #1 livestreaming expert on social media.

Ross is the author and founder of the *100 Predictions* franchise. His book series, *100 Livestreaming & Digital Media Predictions*, reached #1 Amazon Best Seller status in six countries and won 25 book awards. He's been featured by Nasdaq TV, Forbes and Psychology Today.

Relevant Links:

- Website – livestreamuniverse.com
- YouTube Channel – youtube.com/@livestreamuniverse
- Podcast – rossbrandrecordings.com
- Twitter – twitter.com/irossbrand
- Instagram – instagram.com/rossbrand1
- LinkedIn – linkedin.com/in/rossbrand
- Facebook Profile – facebook.com/rossbrandlive
- Facebook Page – facebook.com/livestreamuniverse
- Facebook Group – facebook.com/groups/livestreamuniversecommunity
- Pinterest – pinterest.com/rossbrand
- TikTok – tiktok.com/@rossbrandlive

www.ingramcontent.com/pod-product-compliance
Lightning Source LLC
Chambersburg PA
CBHW071106050326
40690CB00008B/1131